To the Bean Fc

Hugs all Round.

Grandpa Bean (himself)

xxx

Toward Engineering Design Principles for HCI

Synthesis Lectures on Human-Centered Informatics

Editor

John M. Carroll, *Penn State University*

Human-Centered Informatics (HCI) is the intersection of the cultural, the social, the cognitive, and the aesthetic with computing and information technology. It encompasses a huge range of issues, theories, technologies, designs, tools, environments, and human experiences in knowledge work, recreation and leisure activity, teaching and learning, and the potpourri of everyday life. The series publishes state-of-the-art syntheses, case studies, and tutorials in key areas. It shares the focus of leading international conferences in HCI.

Toward Engineering Design Principles for HCI
John Long, Steve Cummaford, and Adam Stork

HCI Design Knowledge: Critique, Challenge, and a Way Forward
John Long, Steve Cummaford, and Adam Stork

Disability Interactions: Creating Inclusive Innovations
Catherine Holloway and Giulia Barbareschi

Participatory Design
Susanne Bødker, Christian Dindler, Ole S. Iversen, and Rachel C. Smith

The Trouble With Sharing: Interpersonal Challenges in Peer-to-Peer Exchange
Airi Lampinen

Interface for an App—The Design Rationale Leading to an App that Allows Someone With Type 1 Diabetes to Self-Manage Their Condition
Bob Spence

Organizational Implementation: The Design in Use of Information Systems
Morten Hertzum

Data-Driven Personas
Bernard J. Jansen, Joni Salminen, Soon-gyo Jung, and Kathleen Guan

Worth-Focused Design, Book 2: Approaches, Context, and Case Studies
Gilbert Cockton

Worth-Focused Design, Book 1: Balance, Integration, and Generosity
Gilbert Cockton

Statistics for HCI: Making Sense of Quantitative
Alan Dix

Usability Testing
Morten Hertzum

Geographical Design: Spatial Cognition and Geographical Information Science, Second Edition
Stephen C. Hirtle

Human-Computer Interactions in Museums
Eva Hornecker and Luigina Ciolfi

Encounters with HCI Pioneers: A Personal History and Photo Journal
Ben Shneiderman

Social Media and Civic Engagement: History, Theory, and Practice
Scott P. Robertson

The Art of Interaction: What HCI Can Learn from Interactive Art
Ernest Edmonds

Representation, Inclusion, and Innovation: Multidisciplinary Explorations
Clayton Lewis

Research in the Wild
Yvonne Rogers and Paul Marshall

Designing for Gesture and Tangible Interaction
Mary Lou Maher and Lina Lee

From Tool to Partner: The Evolution of Human-Computer Interaction
Jonathan Grudin

Qualitative HCI Research: Going behind the Scenes
Ann Blandford, Dominic Furniss, and Stephann Makri

Learner-Centred Design of Computing Education: Research on Computing for Everyone
Mark Guzdial

The Envisionment and Discovery Collaboratory (EDC): Explorations in Human-Centred Informatics with Tabletop Computing Environments
Ernesto G. Arias, Hal Eden, and Gerhard Fischer

Humanistic HCI
Jeffrey Bardzell and Shaowen Bardzell

The Paradigm Shift to Multimodality in Contemporary Computer Interfaces
Sharon Oviatt and Philip R. Cohen

Multitasking in the Digital Age
Gloria Mark

The Design of Implicit Interactions
Wendy Ju

Core-Task Design: A Practice-Theory Approach to Human Factors
Leena Norros, Paula Savioja, and Hanna Koskinen

An Anthropology of Services: Toward a Practice Approach to Designing Services
Jeanette Blomberg and Chuck Darrah

Proxemic Interactions: From Theory to Practice
Nicolai Marquardt and Saul Greenberg

Contextual Design: Evolved
Karen Holtzblatt and Hugh Beyer

Constructing Knowledge Art: An Experiential Perspective on Crafting Participatory Representations
Al Selvin and Simon Buckingham Shum

Spaces of Interaction, Places for Experience
David Benyon

Mobile Interactions in Context: A Designerly Way Toward Digital Ecology
Jesper Kjeldskov

Working Together Apart: Collaboration over the Internet
Judith S. Olson and Gary M. Olson

Surface Computing and Collaborative Analysis Work
Judith Brown, Jeff Wilson, Stevenson Gossage, Chris Hack, and Robert Biddle

How We Cope with Digital Technology
Phil Turner

Translating Euclid: Designing a Human-Centred Mathematics
Gerry Stahl

Adaptive Interaction: A Utility Maximisation Approach to Understanding Human Interaction with Technology
Stephen J. Payne and Andrew Howes

Making Claims: Knowledge Design, Capture, and Sharing in HCI
D. Scott McCrickard

HCI Theory: Classical, Modern, and Contemporary
Yvonne Rogers

Activity Theory in HCI: Fundamentals and Reflections
Victor Kaptelinin and Bonnie Nardi

Conceptual Models: Core to Good Design
Jeff Johnson and Austin Henderson

Geographical Design: Spatial Cognition and Geographical Information Science
Stephen C. Hirtle

User-Centred Agile Methods
Hugh Beyer

Experience-Centred Design: Designers, Users, and Communities in Dialogue
Peter Wright and John McCarthy

Experience Design: Technology for All the Right Reasons
Marc Hassenzahl

Designing and Evaluating Usable Technology in Industrial Research: Three Case Studies
Clare-Marie Karat and John Karat

Interacting with Information
Ann Blandford and Simon Attfield

Designing for User Engagement: Aesthetic and Attractive User Interfaces
Alistair Sutcliffe

Context-Aware Mobile Computing: Affordances of Space, Social Awareness, and Social Influence
Geri Gay

Studies of Work and the Workplace in HCI: Concepts and Techniques
Graham Button and Wes Sharrock

Semiotic Engineering Methods for Scientific Research in HCI
Clarisse Sieckenius de Souza and Carla Faria Leitão

Common Ground in Electronically Mediated Conversation
Andrew Monk

Copyright © 2022 by Morgan & Claypool

All rights reserved. No part of this publication may be reproduced, stored in a retrieval system, or transmitted in any form or by any means—electronic, mechanical, photocopy, recording, or any other except for brief quotations in printed reviews, without the prior permission of the publisher.

Toward Engineering Design Principles for HCI
John Long, Steve Cummaford, Adam Stork

www.morganclaypool.com

ISBN: 9781636393506 Paperback
ISBN: 9781636393513 PDF
ISBN: 9781636393520 Hardcover

DOI 10.2200/S01172ED1V02Y202202HCI055

A Publication in the Morgan & Claypool Publishers series
SYNTHESIS LECTURES ON HUMAN-CENTERED INFORMATICS

Lecture #55
Series Editor: John M. Carroll, Penn State University

Series ISSN 1946-7680 Print 1946-7699 Electronic

Toward Engineering Design Principles for HCI

John Long
University College, London

Steve Cummaford
Ted Baker

Adam Stork
Concerto

SYNTHESIS LECTURES ON HUMAN-CENTERED INFORMATICS #55

ABSTRACT

This is the second of two books by the authors about engineering design principles for human-computer interaction (HCI-EDPs). The books report research that takes an HCI engineering discipline approach to acquiring initial such principles. Together, they identify best-practice HCI design knowledge for acquiring HCI-EDPs. This book specifically reports two case studies of the acquisition of initial such principles in the domains of domestic energy planning and control and business-to-consumer electronic commerce. The book begins by summarising the earlier volume, sufficient for readers to understand the case studies reported in full here.

The themes, concepts, and ideas developed in both books concern HCI design knowledge, a critique thereof, and the related challenge. The latter is expressed as the need for HCI design knowledge to increase its fitness-for-purpose to support HCI design practice more effectively. HCI-EDPs are proposed here as one response to that challenge, and the book presents case studies of the acquisition of initial HCI-EDPs, including an introduction; two development cycles; and presentation and assessment for each. Carry forward of the HCI-EDP progress is also identified.

The book adopts a discipline approach framework for HCI and an HCI engineering discipline framework for HCI-EDPs. These approaches afford design knowledge that supports "specify then implement" design practices. Acquisition of the initial EDPs apply current best-practice design knowledge in the form of "specify, implement, test, and iterate" design practices. This can be used similarly to acquire new HCI-EDPs. Strategies for developing HCI-EDPs are proposed together with conceptions of human-computer systems, required for conceptualisation and operationalisation of their associated design problems and design solutions.

This book is primarily for postgraduate students and young researchers wishing to develop further the idea of HCI-EDPs and other more reliable HCI design knowledge. It is structured to support both the understanding and the operationalisation of HCI-EDPs, as required for their acquisition, their long-term potential contribution to HCI design knowledge, and their ultimate application to design practice.

KEYWORDS

HCI design principles; HCI-EDPs; discipline; engineering; HCI engineering discipline; design knowledge; critique; challenge; design best-practice; domestic energy planning and control; business-to-consumer electronic commerce; domain; case study

Contents

Preface . . . . **xxi**
About This Book xxi
Rationale xxi
About the Authors xxii
About the Readership xxii

Acknowledgments . . . . **xxiii**

Dedication . . . . **xxv**

Terminology . . . . **xxvii**

1 HCI Design Knowledge: Critique, Challenge, and a Way Forward **1**
Summary 1
1.1 HCI 1
1.2 HCI Engineering 1
1.3 HCI Engineering Design 2
1.4 HCI Engineering Design Principles 2
    1.4.1 To Support "Specify, Implement, Test, and Iterate" HCI Design 3
    1.4.2 To Support "Specify then Implement" HCI Design 4
1.5 Critique, Challenge, and a Way Forward 5
Review 6
1.6 Practice Assignment 6

**2 Introduction to Initial HCI Engineering Design Principles for Domestic
Energy Planning and Control** **9**
Summary 9
2.1 Conception of Substantive HCI Engineering Design Principles 9
    2.1.1 Dowell and Long's Conception of the General HCI Engineering
           Design Problem for an Engineering Discipline of HCI 9
    2.1.2 Conception of Substantive HCI Engineering Design Principles . . . 10
    2.1.3 Conceptions of the General HCI Design Problem and General
           HCI Design Solution 12

	2.1.4	Conceptions of the Specific HCI Design Problem and the Specific HCI Design Solution	15
	2.1.5	Conception of Substantive HCI Engineering Design Principles Revisited	15
2.2		Strategy for Developing HCI Engineering Design Principles	15
	2.2.1	Strategy Development	16
	2.2.2	Comparison with Alternative Strategies	18
	2.2.3	Scoping the Research Using the Potential for Planning and Control HCI Engineering Design Principles	18
	2.2.4	Acquiring Potential Guarantee	19
	2.2.5	Shorter-Term Research Benefits	19
	2.2.6	Overview of MUSE	19
	2.2.7	Development Cycle User Requirements Selection Criteria	21
2.3		Conception of Human-Computer Systems	24
	2.3.1	Interactive Worksystem Costs	24
	2.3.2	Potential Human Cognitive Structures	25
	2.3.3	Potential Human Physical Structures	26
	2.3.4	Potential Computer Abstract Structures	26
	2.3.5	Potential Computer Physical Structures	26
Review			27
2.4		Practice Assignment	27
	2.4.1	General	27
	2.4.2	Practice Scenarios	28

3 Cycle 1 Development of Initial HCI Engineering Design Principles for Domestic Energy Planning and Control **29**

Summary			29
3.1		Operationalising Specific Design Problems and Solutions	29
	3.1.1	Framework for Task Quality	29
	3.1.2	Framework for Interactive Worksystem Costs	30
	3.1.3	Composite Structures	31
3.2		Conception of Planning and Control	33
	3.2.1	Conceptions of Planning and Control Claiming Design Guidance	33
	3.2.2	Conceptions of Planning and Control with No Claims for Design Guidance	35
	3.2.3	Initial Conception of Planning and Control	36

		3.2.4	Operationalisation	41
	3.3	Cycle 1 Best-Practice Development		41
		3.3.1	User Requirements	42
		3.3.2	Artefact Specification	42
		3.3.3	Best-Practice Development	43
		3.3.4	Evaluation	45
	3.4	Cycle 1 Operationalisation		46
		3.4.1	Current Solution Operationalisation	47
		3.4.2	Specific Design Problem Operationalisation	49
		3.4.3	Specific Design Solution Operationalisation	50
	Review			52
	3.5	Practice Assignment		52
		3.5.1	General	52
		3.5.2	Practice Scenario	53

4 Cycle 2 Development of Initial HCI Engineering Design Principles for Domestic Energy Planning and Control. **55**

	Summary			55
	4.1	Cycle 2 Best-Practice Development		55
		4.1.1	User Requirements	55
		4.1.2	Artefact Specification	56
		4.1.3	Best Practice Development	56
	4.2	Cycle 2 Operationalisation		59
		4.2.1	Generality Concern	59
		4.2.2	Current Solution Operationalisation	59
		4.2.3	Specific Design Problem Operationalisation	62
		4.2.4	Specific Design Solution Operationalisation	62
	Review			65
	4.3	Practice Assignment		65
		4.3.1	General	65
		4.3.2	Practice Scenario	65

5 Initial HCI Engineering Design Principles for Domestic Energy Planning and Control. **67**

	Summary			67
	5.1	Detailed Strategy		67
		5.1.1	Generality of the Initial HCI Engineering Design Principles	68
		5.1.2	Generalisation over Types	69

xiv

5.2 Initial HCI Engineering Design Principles Identified During
Operationalisation(s) 69
    5.2.1 Examples 69
5.3 Initial Assumption Assessment from Operationalisation(s) 72
    5.3.1 Examples 72
5.4 Inspirational Initial HCI Engineering Design Principles from
Operationalisation(s) 73
    5.4.1 Examples 73
5.5 Initial HCI Engineering Design Principles from General Guidelines 77
    5.5.1 Examples 77
5.6 Initial HCI Engineering Design Principles from MUSE Guidelines 79
    5.6.1 Example 79
5.7 Initial HCI Engineering Design Principles from MUSE Tasks 80
    5.7.1 Example 80
Review 81
5.8 Practice Assignment 81
    5.8.1 General 81

**6 Assessment and Discussion of Initial HCI Engineering Design Principles
for Domestic Energy Planning and Control............................. 83**
Summary 83
6.1 Strategy Assessment and Discussion 83
    6.1.1 Strategy and Conception Changes 83
    6.1.2 Status of Initial HCI Engineering Design Principles 84
    6.1.3 Strategy Assessment 84
    6.1.4 Further Research 84
    6.1.5 Further Strategy Discussion 85
6.2 MUSE for Research (MUSE/R) 86
    6.2.1 Scope and Notation 86
    6.2.2 Process 87
    6.2.3 Support for Design 88
    6.2.4 Further Research 88
Review 88
6.3 Practice Assignment 88
    6.3.1 General 88
    6.3.2 Practice Scenarios 89

xv

7   Introduction to Initial HCI Engineering Design Principles for Business-to-Consumer Electronic Commerce **91**

Summary 91

7.1   Conception of HCI Engineering Design Principles 91

    7.1.1   Introduction 91

    7.1.2   Conception of the General HCI Design Problem 92

    7.1.3   Conception of the General HCI Design Solution 92

    7.1.4   Conception of the General HCI Engineering Design Principle 93

7.2   Strategy for Developing HCI Engineering Design Principles 96

    7.2.1   Introduction 96

    7.2.2   Instance-First and Class-First Strategies 96

    7.2.3   HCI Engineering Design Principles as Class Design Knowledge ... 97

    7.2.4   Class Development 97

    7.2.5   Definition of Classes 98

    7.2.6   Conception of Classes of Design Problem 98

    7.2.7   Conception of Classes of Design Solution 98

    7.2.8   Identification of Promising Classes 99

7.3   Method for Operationalising the Class-First Strategy 100

    7.3.1   Introduction 100

    7.3.2   Specification Method for Class Design Problem and Class Design Solution 100

    7.3.3   Specification Method for HCI Engineering Design Principles .. 103

7.4   Identification of Class Design Problems 105

    7.4.1   Introduction 105

    7.4.2   Selection of Potential Class Design Problem 105

    7.4.3   Class of Design Problem for Transaction Systems 107

    7.4.4   Specification of Sub-Classes for Transaction Systems 109

Review 112

7.5   Practice Assignment 112

    7.5.1   General 112

    7.5.2   Practice Scenarios 114

8   Cycle 1 Development of Initial HCI Engineering Design Principles for Business-to-Consumer Electronic Commerce........................... **115**

Summary 115

8.1   Cycle 1 Development 115

    8.1.1   Introduction 115

xvi

	8.1.2	Selection of Systems for Specific Design Problem and Specific Design Solution Development	116
	8.1.3	Testing Procedure	116
	8.1.4	Specify Specific Design Problems	119
	8.1.5	Specify Class Design Problem	121
	8.1.6	Evaluate Class Design Problem	122
	8.1.7	Specify Class Design Solution	122
	8.1.8	Specify Specific Design Solutions	122
	8.1.9	Evaluate Class Design Solution	123
8.2		Cycle 1 Class Design Problem/Class Design Solution Specification	124
	8.2.1	Introduction	124
	8.2.2	Stage 1: Specify Specific Design Problems	124
	8.2.3	Stage 2: Specify Class Design Problem	131
	8.2.4	Stage 3: Evaluate Class Design Problem	138
	8.2.5	Stage 4: Specify Class Design Solution	139
	8.2.6	Stage 6: Evaluate Class Design Solution	145
Review			147
8.3		Practice Assignment	147
	8.3.1	General	147
	8.3.2	Practice Scenarios	148

9 Cycle 2 Development of Initial HCI Engineering Design Principles for Business-to-Consumer Electronic Commerce 149

Summary			149
9.1		Cycle 2 Development	149
	9.1.1	Introduction	149
	9.1.2	Selection of Systems for Specific Design Problem and Specific Design Solution Development	150
	9.1.3	Testing Procedure	150
	9.1.4	Specify Specific Design Problems	150
	9.1.5	Specify Class Design Problem	150
	9.1.6	Evaluate Class Design Problem	150
	9.1.7	Specify Class Design Solution	150
	9.1.8	Specify Specific Design Solutions	150
	9.1.9	Evaluate Class Design Solution	150
9.2		Cycle 2 Class Design Problem/Class Design Solution Specification	150
	9.2.1	Introduction	151
	9.2.2	Stage 1: Specify Specific Design Problems	151

		9.2.3	Stage 2: Specify Class Design Problem	151
		9.2.4	Stage 3: Evaluate Class Design Problem	151
		9.2.5	Stage 4: Specify Class Design Solution	151
		9.2.6	Stage 6: Evaluate Class Design Solution	152
	9.3	Practice Assignment		152
		9.3.1	General	152
		9.3.2	Practice Scenarios	153

10 Initial HCI Engineering Design Principles for Business-to-Consumer Electronic Commerce 155

Summary 155

10.1 HCI Engineering Design Principle Specification Requirements 155

10.1.1 HCI Engineering Design Principle Specification Method 155

10.1.2 HCI Engineering Design Principle Components 155

10.1.3 HCI Engineering Design Principle Scope 156

10.1.4 HCI Engineering Design Principle Specification 156

10.1.5 HCI Engineering Design Principle Achievable Performance 156

10.2 HCI Engineering Design Principles Acquired in Cycle 1 Development . . 156

10.2.1 HCI Engineering Design Principle Scope 157

10.2.2 HCI Engineering Design Principle Specification 159

10.2.3 HCI Engineering Design Principle Achievable Performance 163

10.3 HCI Engineering Design Principles Acquired in Cycle 2 Development. . . 163

10.3.1 HCI Engineering Design Principle Scope 163

10.3.2 HCI Engineering Design Principle Specification 167

10.3.3 HCI Engineering Design Principle Achievable Performance 171

10.4 Initial HCI Engineering Design Principles 171

Review 172

10.5 Practice Assignment 172

10.5.1 General 172

10.5.2 Practice Scenario 173

11 Assessment and Discussion of Initial HCI Engineering Design Principles for Business-to-Consumer Electronic Commerce 175

Summary 175

11.1 Introduction 175

11.2 Strategy Assessment 175

11.3 Discussion 175

11.3.1 Specific Design Problems Specification 176

	11.3.2	Class Design Problem Specification	178
	11.3.3	Class Design Problem Evaluation	178
	11.3.4	Class Design Solution Specification	178
	11.3.5	Specific Design Problem Specification	178
	11.3.6	Class Design Solution Evaluation	179
	11.3.7	Class Design Problem to Class Design Solution Mapping	179
	11.3.8	HCI Engineering Design Principle Definition Method	179
	11.3.9	Initial HCI Engineering Design Principles	179
	11.3.10	Requirement for Validation, Leading to Guarantee Review	180
	11.3.11	HCI Conceptions Review	180
	11.3.12	HCI Engineering Design Principle Applicability and Potential as Design Support	180
	11.3.13	Future Research Discussion	181
11.4		Business-to-Consumer Best Practice Update	181
Review			182
11.5		Practice Assignment	183
	11.5.1	General	183
	11.5.2	Practice Scenarios	183

12 Progress in Carry Forward of HCI Engineering Design Principles for Future Research **185**

Summary			185
12.1		Toward HCI Engineering Design Principles - General Progress and Carry Forward	185
	12.1.1	Domestic Energy Planning and Control	185
	12.1.2	Business-to-Consumer Electronic Commerce	186
	12.1.3	Research Progress	187
12.2		HCI Engineering Design Principles: Research Remaining	187
	12.2.1	Domestic Energy Planning and Control	188
	12.2.2	Business-to-Consumer Electronic Commerce	188
	12.2.3	Research Remaining	188
Review			189
12.3		Practice Assignment	189
	12.3.1	General	189
	12.3.2	Practice Scenarios	192

Postscript **193**

Bibliography **195**

Authors' Biographies **201**

Preface

ABOUT THIS BOOK

This book is one of two. The companion volume is *HCI Design Knowledge - Critique, Challenge, and a Way Forward* (Long et al., 2022, in press). The title of the present book describes its scope and content. The scope is HCI (human–computer interaction) and HCI engineering design principles (HCI-EDPs). The content is the acquisition of initial HCI-EDPs. The challenge to existing HCI design knowledge is identified as its lack of reliability, when applied to design practice. Its support for the latter, then, needs to be more effective. HCI-EDPs are one response to this challenge.

The book presents "instance-first" and "class-first" approaches to the acquisition of HCI-EDPs, as instantiated in the two case studies. The EDPs are constructed from solutions to design problems, themselves derived from user requirements. The case study application domains are domestic energy planning and control, and business-to-consumer electronic commerce. Both report the acquisition of initial HCI-EDPs and comprise chapters on: their introduction; two cycles of development; and their presentation and assessment.

The publication of such a book is timely. Both approaches to the acquisition of HCI-EDPs are novel. They have little by way of competition with the exception of "design patterns." A review of the HCI research literature indicates the need for more effective support for HCI design practice. The case studies espouse the same HCI discipline and HCI engineering conceptions. Any differences in operationalisation inform future research, for example, the strategy adopted and the type of principle acquired. Both case studies serve to support researchers to build on the work. Practice assignments at the end of each chapter offer further support for research applications. The final chapter suggests carry forward to support future research of acquiring HCI-EDPs.

RATIONALE

Such a book is needed. The strategies for acquiring HCI-EDPs constitute a novel contribution to HCI research. Further, the book differs from other books concerning HCI design knowledge and its application, such as Ritter, Baxter, and Churchill (2014)'s *Foundations for Designing User-Centered Systems*; Hartson and Pyla (2018)'s *The UX Book: Agile UX Design for a Quality User Experience*; Kim (2020)'s *Human-Computer Interaction - Foundations and Practice*; and Zagalo (2020)'s *Engagement Design: Designing for Interaction Motivations*. In contrast, the present book conceives HCI design knowledge in terms of an HCI engineering discipline. HCI-EDPs support "specify then implement" design practice. They are acquired by the existing best practice at the time of their acquisition. The book constitutes a contrast with the books referenced. Their best practice, however, can be applied currently to acquire additional HCI-EDPs.

The book also differs from Long (2021)—*Approaches and Frameworks for HCI Research*. He proposes an approach and a framework for HCI as engineering and also refers to specific and general HCI engineering principles. He underlines the need for empirical validation to increase the reliability of HCI design knowledge. However, unlike the present book, no conception or operationalisation of HCI-EDPs is offered. Last, Long presents no case studies as reported here. In short, this book is best understood as starting from where Long left off.

ABOUT THE AUTHORS

The authors feel qualified to write such a book. They developed the approach to HCI-EDP acquisition during their time at University College London. It has been used to frame HCI research teaching and to support associated research of which the two case studies are an example.

Chapters 2–6 are based on Stork's Ph.D. thesis (1999) and Chapters 7–11 on that of Cummaford (2007). Comparable work has yet to be published elsewhere with the exception of "design patterns." Long was the initiator of the work and supervisor in both cases. He is also responsible with Springer for bringing the book to publication. For these reasons, Long appears as first author, with Cummaford and Stork ordered alphabetically. However, all chapters have been reviewed by all authors.

Last but not least, the book would not have been possible without the associated research and support of colleagues, especially that of John Dowell, as well as of Ph.D. students at the EU/UCL Unit (University College London).

ABOUT THE READERSHIP

This research book is for graduate and postgraduate students of HCI. It is also for young academic researchers and their supervisors. Practice assignments, at the end of each chapter, support their understanding and application of the concepts presented. In particular, the case study-related assignments support researchers in conceptualising and operationalising HCI-EDPs and so building on the present research in the same and different domains of application. The book is also of interest to researchers in related disciplines and movements, contributing to HCI, such as cognitive psychology, UX-design, design science, software engineering, design research, human-factors, agile design, cognitive ergonomics, and human-centred informatics (see earlier references).

Acknowledgments

This book is offered as a tribute to colleagues and Ph.D. students at the EU/UCL Unit (University College London), whose earlier research contributions have made it possible (see also <hciengineering.net> for more information concerning those contributions).

Thanks are due to Jack Carroll for including the book in his Synthesis Lectures on Human-Centred Informatics Series as a "red bloodied" (sic) research outlier intended to maintain the series coverage. Also, to Diane Cerra, Christine Kiilerich, and Deborah Gabriel for their enthusiastic and faultless efforts in bringing the book to print. Thanks are also due to anonymous reviewers, who have contributed to improving the clarity and coverage of an earlier draft.

Dedication

To our families and friends for their support, patience, and understanding.

Terminology

AB Testing
Agile Methods
Air Traffic Management
   Planning Horizon
   Theory of Planning Horizon (TOPH)
Apple
Applied Psychology
Architecture
   Computer
   Human
Artificial Intelligence (AI)
Atomic Design Methods
Best-Practice Design
   Cycle 1 Development
   Cycle 2 Development
Best Selling Lists
Betty Brown Teapot
Black Tea
Boolean Values
Civil Engineering
Cognitive Engineering
Cognitive Ergonomics
Cognitive Psychology
Computer
   Model
   Representation States
   Representation Structure States
Computer Costs
   Abstract
      Behavioural
      Structural

   Physical
      Behavioural
      Structural
Computing Technology
Conceptions
   Actual
      Costs
      Performance
      Quality
   Changes
   Cognitive Structures
   Control
      Domain
      Worksystem
   Desired
      Costs
      Performance
      Quality
   Planning
      Domain
      Worksystem
Costs Matrix
Design
   Craft Artefacts
   Guidance
   Knowledge
   Practice Experience
   Rationale
   Scenario-based
   Testing
      AB

Funnel
Design Patterns
Direct Manipulation Theory
Domain
   Boundary Criteria
   Diagram Key
   Model
      Class
      Instance
Domestic Central Heating
   Case Study
   Planning and Control
Ease of Use
Ecological Theory
E-Commerce
Effective Support
Electronic Commerce
   Case Study
   Mercantile Models
Emergency Management Coordination Response System (EMCRS)
Engineering
   Electronic
Engineering Principles
Ergonomics
E-Shops
External Cognition Theory
Face Recognition
FeedFinder
Firefox
Formality
Formal Knowledge
Frameworks
   Interactive Worksystem Costs
   Task Quality
Funnel Testing

Gaming
   Guarantees
   Procedural
Goals, Operators, Methods, Strategies (GOMS)
Goods
   Physical
   Informational
Google Chrome
Graphical User Interface (GUI)
Grounded Theory
Guidelines
   Electronic Commerce
   Domestic Energy
   General
   MUSE
Heuristics
HCI
   Design Problem
      Class
      General
      Specific
   Design Solution
      Class
      General
      Specific
   Human-Computer Systems
   Planning and Control
HCI Engineering Design Principles (HCI-EDPs)
   Achievable Performance
   Acquisition Approach
      Class-first
      Instance First
   Classification Space
   Declarative

Generality
General Guidelines
Inspirational
MUSE Guidelines
MUSE Tasks
Procedural
State Stream
HCI Design
  Challenge
  Critique
  Practice
    Best-Practice
    Specify and Implement
    Specify, Implement and Test
    Specify, Implement, Test and Iterate
    Specify then Implement
    Scenario-based
  Way Forward
HCI Design Knowledge
  Challenge
  Critique
  Declarative
  Design Rationale
  Prescriptive
  Procedural
HCI Design Patterns
HCI Engineering
  Classification Space
  Components
  Design
    Problem
    Solution
  Principles
HCI Heuristics
HCI Knowledge
HCI Models and Methods

HCI Particular Scope
  Computers
  Humans
  Interactions
  Performance
HCI Principles
HCI Progress
  Electronic Commerce
  Domestic Energy
  General
HCI Research
  Craft Artefacts
  Designer Experience
  Heuristics
  Models and methods
  Principles
  Progress
  Rules
  Validation
HCI Rules
HCI Validation
Herbs of Grace
Heuristics
Hierarchy of Complexity
Human Architecture
Human-Centred Informatics
Human-Computer Interaction (HCI)
Human-Computer Systems
Human Costs
  Abstract
    Behavioural
    Structural
  Physical
    Behavioural
    Structural
Human Factors

Implementation
Innovation
Interaction Design
Interdisciplinary Overlapping Fields
Internet Explorer
Jackson System Development
Jamster
Knowledge
  Declarative (Substantive)
  Decomposition
  Procedural (Methodological)
  Recomposition
  Solution
Manchester United
  Text Alerts Service
Mercantile Models
Method for Usability Engineering (MUSE)
Method for Minimal Viable Product (MMVP)
Model Human Processor (MHP)
Models
  Resource
Models and Methods
Multi-media
  Challenge
  Critique
Multiple Task Work
MUSE for Research (MUSE/R)
On-line
  Banking
  Dating
  Shopping
Operationalisation
  Cycle 1
    Initial HCI-EDPs
    Selection

Specific Actual Costs
Specific Actual Performance
Specific Actual Quality
Specific Design Problems
Specific Design Solutions
Specific Desired Costs
Specific Desired Quality
Cycle 2
  Initial HCI-EDPs
  Selection
  Specific Actual Costs
  Specific Actual Performance
  Specific Actual Quality
  Specific Design Problems
  Specific Design Solutions
  Specific Desired Costs
  Specific Desired Quality
Performance
  Achievable
  Actual
    Specific
  Desired
  Potential Guarantee
Planning and Control
  Composite
  Desired States and Structures
  Generalised
  Interleaved
  With Design Guidance
  Without Design Guidance
Plans
  In Domain
  In Interactive Worksystem
  In Separate Worksystems
  Worksystem
PowerPoint

Principles
Procedural Knowledge
Prototypes
Rules
Safari
Secretarial Office Administration
Simulation
Smart Phones
Snow Valley
Software Engineering
Specification
Speed and Errors
Standards
Stash Tea
Storyboard Scenarios
Strategy
  Assessment
  Changes
   Developing Engineering Design Principles
    Detailed Strategy
     Bottom Up
     Middle-Out
     Top Down
  Discussion
Streams
  Behaviour
  Structure
Structured Software Analysis and Design Methods (SSADM)
Structures
  Composite
Supercraft
System Versions
Task Quality
  Actual Quality

Specific
Desired Quality
Testing Procedure
  Setup
  Participants
  Tasks
Theory
  Of Operator Planning Horizons (TOPH)
Trial and Error
Usability
User
  Affective
  Cognitive
  Conative
  Model
    Class Model
    Instance Model
  Representation States
  Representation Structure States
User Costs
  Actual
    Specific
  Calculation
    Abstract Behavioural (Mental)
     Desired
      Human
      Structural
    Physical Behavioural
     Structural
  Costs Matrix
User Experience (UX)
  UX design
User Requirements
  Identification
  Selection Criteria
Validation

Widgets

Wire-Frame Models

Worksystem

    Boundary Criteria

    Class

World Health Organisation – WHO

World of Work

    Desired Performance

    Desired Quality

    Attributes

        Abstract

        Physical

        Related

        Task Goals

        Unrelated

    Objects

    Product Goals

        Achieved

    Task Goals

        Achieved

    States

    Structures

Xerox

Yourdon

CHAPTER 1

HCI Design Knowledge: Critique, Challenge, and a Way Forward

SUMMARY

This chapter introduces HCI design knowledge, together with an associated critique, challenge and a way forward. This chapter defines: Human-Computer Interaction (HCI); HCI engineering; HCI engineering design, and HCI engineering design principles (HCI-EDPs). The latter are acquired by applying "best-practice" design knowledge at the time of their acquisition, using "specify, implement, test, and iterate" design practice. HCI-EDPs support both "specify and implement" and "specify then implement" HCI design practice, depending on the completeness of their design problem specification. Design problems are derived from user and any other requirements.

1.1 HCI

HCI is conceived here as a discipline. This conception provides a framework for HCI. As a discipline, HCI is constituted of an HCI general problem, with a particular scope, which conducts research (Long and Dowell, 1989). The HCI general problem is one of design (Dowell and Long, 1989). The particular scope of the general problem of HCI design is—"humans interacting with computers to do something as desired" (see Long, 2021). HCI research acquires and validates knowledge to support practices in solving the general HCI problem of design within its particular scope of the HCI Discipline

1.2 HCI ENGINEERING

HCI, as a discipline, is conceived in different ways, for example: innovation; art; craft; applied; engineering; and science (Long, 2021). HCI is conceived here as an engineering discipline, constituted of an HCI engineering general problem with an HCI engineering particular scope and which conducts HCI engineering research.

The HCI general engineering problem is one of HCI design for performance. The particular scope of the general problem of HCI engineering design is humans interacting with computers to do something as desired. HCI engineering research acquires and validates explicit design knowledge to support HCI practices of design. The explicit HCI engineering knowledge solves the general HCI engineering problem of design for performance with the particular scope of humans interacting with computers to do something as desired.

1.3 HCI ENGINEERING DESIGN

HCI engineering design is conceived here as the specification and implementation of interactive human-computer systems for performance, which do something as desired. The HCI engineering design knowledge supports specification and implementation design practices. The design knowledge comprises declarative (that is, substantive) knowledge, the "what" of design and the procedural (that is, methodological) knowledge, and the "how" of design. Specification conceptualises human-computer interactive systems, and implementation operationalises them in the form of an actual or potential artefact, satisfying user and any other requirements. Together, they solve the HCI general problem of design. Their effectiveness is determined by the reliability of the support they receive from HCI design knowledge, acquired by HCI engineering research.

1.4 HCI ENGINEERING DESIGN PRINCIPLES

According to the HCI research literature, HCI principles, interpreted broadly, constitute a common form of design knowledge (see also Long, 2021). Such principles may be expressed as truth proposition, law, heuristic, rule, guideline, diagnosis and prescription, doctrine or theory, as required by HCI research and practice. However, in all cases, HCI design principles serve as the object of HCI research, as the acquisition and validation of HCI design knowledge. They embody the key purpose of HCI design knowledge, which is to support HCI design practice.

However, with notable exceptions (Card, Moran, and Newell, 1983), a review of the HCI literature fails to identify any HCI-EDPs, which, although created, have been explicitly operationalised or validated, as required by engineering HCI knowledge to ensure its effectiveness in the support of design practice. Indeed, it is difficult to see how many of the claimed principles could be validated explicitly. Although conceptualised, the principles lack the required completeness and coherence to be operationalised for their explicit acquisition. They cannot, then, be tested or generalised. The failure, however, does not result from the absence of references to HCI principles in the literature. Such references, interpreted broadly, include: for example, Wickens (1984, 1993), Wickens, Lee, and Becker (2004), and Norman (2013) as principles; Shneiderman (1983, 2010) as rule; and Nielsen (1993) as heuristics. It is rather due to the nature and state of the principles, proposed as HCI knowledge. It is argued here that the latter fail to support design practice effectively.

1.4.1 TO SUPPORT "SPECIFY, IMPLEMENT, TEST, AND ITERATE" HCI DESIGN

HCI design practices can be categorised in different ways. For current purposes, the categorisation of Dowell and Long (1989) is adopted. "Specify, implement, test, and iterate" HCI design practice is supported by "informal design knowledge," which requires empirical test to ensure its effective application. Illustration of such informal design knowledge, in the form of HCI models and methods and HCI principles, rules and heuristics follows.

1.4.1.1 HCI Models and Methods

HCI models and methods pervade the research literature.

Concerning HCI models, the "model human processor" is probably the best-known model of this sort (Card et al., 1983), contributing to their psychology framework for HCI. Barnard (1991), as part of his cognitive psychology-bridging framework, includes a family of cognitive task models, in addition to the foundational interacting cognitive subsystems model. Carroll (2003) proposes models associated with scenario-based design and the associated design rationale. These models form part of HCI frameworks.

HCI theories, however, also reference models. Wright, Fields, and Harrison (2000) propose a resource model, as part of external cognition theory. Kirsh (2001) proposes a model of "entry points," as part of ecological theory.

As concerns methods, they can also be exemplified by: "grounded theory" of Glaser and Strauss (1967, and one of the earliest); "direct manipulation" of Shneiderman (1983, and one of the best known); "MUSE Method for USability Engineering" (Lim and Long, 1994); "multidisciplinary practice in requirements engineering" (MDP/RE) of Denley and Long (2001); and risk-related methods, such as Boehm and Lane (2006) and Pew and Mavor (2007).

Concerning HCI models and methods together, both are addressed in the work of Dowell (1998), as a model of cognitive design formulation and as an associated method. Also, of Rauterberg (2006) as a model of the interaction space and a method for validation. Further exemplification is offered by Carroll (2003, 2010). The model is associated with design rationale and the method with scenario-based design. Also, by Hill (2010) in the domain of the co-ordination of the emergency services in response to disasters (EMCRS). The models include those of the EMCRS interactive system and its domain of application. The method applies the models to the diagnosis of design problems and to suggest the prescription of design solutions.

4   1. HCI DESIGN KNOWLEDGE: CRITIQUE, CHALLENGE, AND A WAY FORWARD

1.4.1.2 HCI Principles, Rules and Heuristics

The earliest HCI design principles are probably those of Wickens (1984) for display design. They include the "principle of pictorial realism" and the "principle of avoiding absolute judgement limits." The principles of Norman (1983, 1989, 1993, and 2013) are for HCI design. They include the "principle of getting the mappings right" and the "principle of exploiting the power of constraints, both natural and artificial."

Perhaps the best-known and most referenced rules for HCI design are the six "golden" rules of Shneiderman (1983,1998, and 2010). They include the "rule of short-term memory load reduction" and the "rule of internal locus of control support."

Nielsen's 10 heuristics (1993) for HCI design include the "heuristic of recognition rather than recall" and the "heuristic of error prevention."

Other comparable types of design knowledge include standards (Bevan, 2001) and directives (Rauterberg and Krueger, 2000).

1.4.2   TO SUPPORT "SPECIFY THEN IMPLEMENT" HCI DESIGN

In contrast to "specify, implement, test, and iteration" design practice (see § 1.4.1) "specify then implement" HCI design practice is supported by "formal design knowledge," as HCI-EDPs (Dowell and Long, 1989). The latter, as design knowledge, do not require empirical test to ensure their reliable application. This is in contrast to satisfying user and any other requirements, which still need "specify, implement, test, and iteration" design practice. Such "specify then implement" HCI-EDPs do not currently exist (if "design patterns" are excluded—but see § 1.4.2.2).

1.4.2.1 HCI Engineering Design Principles

However, Dowell and Long (1989), although not presenting any such HCI-EDPs, propose a detailed theoretical characterisation of them, expressed in the form of a classification space for design disciplines. They argue that HCI-EDPs should form the foundation for future HCI engineering knowledge. Such HCI-EDPs would form prescriptive HCI design knowledge, supporting the solution of "hard" (completely specifiable) design problems for a "specify then implement" design practice. Cummaford and Long (1998) argue further that current HCI design knowledge is insufficiently well specified to be validated. More formal design knowledge, however, would be validatable (Cummaford, 2000). Stork (1999) further proposes a strategy for developing substantive such engineering principles. The "instance-first" strategy involves the identification of general relationships between specific design problems and their solutions (see Chapters 2–6). Further, Cummaford (2000, 2007) proposes a "class-first" strategy that specifies problems and solutions at the level of classes as required for HCI-EDP acquisition (see Chapters 7–11).

1.4.2.2 Design Patterns

It might be argued that "design patterns" (Bayle et al., 1997 and Seffah, 2015) constitute an exception to the general claim that HCI-EDPs do not currently exist. This is a serious claim.. The approach has much in common with that presented here. However, "design patterns" lacks the completeness and coherence afforded by the HCI discipline and HCI engineering conceptions of HCI-EDPs (Stork, 1999). Nevertheless, carrying forward the latter here is considered to include the carrying forward of the former.

1.5   CRITIQUE, CHALLENGE, AND A WAY FORWARD

A critique of the effectiveness of HCI design knowledge to support HCI design practice, as required by an engineering discipline of HCI, finds the former wanting. Ineffectiveness, then, constitutes a challenge for HCI design knowledge. The latter holds for the knowledge generally, whether models and methods; principles, rules, and heuristics; or HCI-EDPs.

For all types of design knowledge once acquired, the challenge can be met by empirical validation. However, such case studies appear only by exception. John and Gray (1995), Atwood, Gray, and John (1996), and Teo and John (2008) report case studies to validate the models of Card et al. (1983). Elsewhere, there are case studies by: Long and Monk (2002) of a conception for telemedical consultation research; Long and Brostoff (2004) for dementia care; Lim and Long (1994) for a structured analysis and design Method for USability Engineering (MUSE); and Long and Hill (2005) for a theory of the operator planning horizon (TOPH) for air traffic management (Timmer and Long, 2002) .

In contrast, HCI-EDPs are proposed here as one way to meet the challenge of the ineffectiveness of design knowledge in addition to empirical validation. Strategies are proposed for their acquisition, such that they be both validatable and validated.

Cummaford and Long (1998) propose a "class-first" strategy, embedding principle generalisation in the acquisition process itself of the HCI-EDP. Generalisation is the final stage of the empirical validation process. Such principles are general to classes of design solutions to classes of design problems. Stork (1999) proposes an "instance-first" strategy for developing such principles. It requires the identification of general relationships between specific design problems and their solutions. Last, Cummaford (2007) proposes that the specification of problems and solutions, at the level of classes, be required in the acquisition of such principles. Iterative identification is needed for both class design problems and their class design solutions. The commonalities between them and between the commonalities themselves form the basis for an HCI-EDP. The latter would then apply to all HCI design problems within its scope. Such formal (that is completely specified) HCI-EDPs would offer the possibility to "specify then implement" design solutions to "hard" (determinate) design problems.

Stork's (1999) research forms the basis for the case study of domestic energy planning and control, reported in Chapters 2–6. Cummaford's research (2007) forms the basis for the case study of business-to-consumer electronic commerce, reported in Chapters 7–11. Together they constitute a way forward towards HCI-EDPs.

REVIEW

The chapter introduces HCI design knowledge, together with an associated critique, a challenge and a way forward. The chapter defines Human-Computer Interaction (HCI), HCI engineering, HCI engineering design, and HCI-EDPs. HCI design knowledge supports either "specify, implement, test, and iterate" HCI design, in the form of models and methods, and principles, rules, and heuristics. Or, it also affords "specify then implement" HCI design, as HCI-EDPs (and possibly "design patterns"). The chapter forms the basis for the acquisition of initial HCI-EDPs in the domains of domestic energy planning and control and of business-to-consumer electronic commerce.

1.6 PRACTICE ASSIGNMENT

Describe the assumptions made by your research as concerns: HCI; HCI Engineering; HCI engineering design, and HCI-EDPs. If you have no research of your own at this time, select the research of a colleague or supervisor. Alternatively, select a suitable publication from the HCI research literature.

- What type of knowledge is your (or other's) research attempting to acquire?

- What type of design practice is the knowledge intended to support?

- What aspects of validation do the research address?

- Contrast the similarities and differences between the assumptions/knowledge/design practice/validation made by your (or other's) research and the proposals made here.

- How might the differences be made coherent? If they cannot be made coherent, why might this be so?

Hints and Tips

Difficult to get started?

Try reading this chapter again, while at the same time thinking about how to describe your own research (or that of others). Note similarities and differences between the two lines of thought as you go along.

1.6 PRACTICE ASSIGNMENT 7

- Describe your research (or that of others) in its own terms before attempting to apply those proposed here.

Difficult to complete?

Familiarise yourself with the main ways of conceptualising HCI, identified in the HCI research literature, before attempting to address those proposed here.

Test

List as many of the section headings as you can from memory.

CHAPTER 2

Introduction to Initial HCI Engineering Design Principles for Domestic Energy Planning and Control

SUMMARY

This chapter introduces HCI-EDPs for the application domain of domestic energy planning and control. The introduction comprises: a conception of declarative EDPs; an "instance-first" strategy for developing such principles, and a conception of human-computer systems, required for such principles. The chapter constitutes the basis for the following chapter on the development of initial HCI-EDPs to support the more effective design of interactive domestic energy planning and control systems.

2.1 CONCEPTION OF SUBSTANTIVE HCI ENGINEERING DESIGN PRINCIPLES

The development of HCI-EDPs is one response to the challenge of increasing the reliability of HCI design knowledge to support design practice more effectively. The case study makes progress towards such HCI-EDPs in the application domain of domestic energy planning and control. The progress constitutes potential carry forward for future HCI-EDP research.

2.1.1 DOWELL AND LONG'S CONCEPTION OF THE GENERAL HCI ENGINEERING DESIGN PROBLEM FOR AN ENGINEERING DISCIPLINE OF HCI

The conception of Dowell and Long (1989) comprises a set of related concepts, expressing the general design problem of an engineering discipline of HCI (Long and Dowell, 1989). HCI-EDPs embody such concepts. The conception, as espoused by the case study, follows.

Dowell and Long conceptualise the general HCI design problem as the design of interactive "worksystems" (sic) for performance. That is, the design of behaviours constituting a worksystem {S}, whose actual performance (Pa) conforms to some desired performance (Pd). To design {S} would require the design of human behaviours {U} interacting with computer behaviours {C}.

Hence, conception of the general design problem of an engineering discipline of HCI is expressed as: specify then implement {U} and {C}, such that {U} interacting with {C} = {S} Pa=Pd, where Pd = fn (Qd, Kd). Qd expresses the desired quality of the products of work for a domain of application and Kd expresses acceptable (that is, desired) costs, incurred by the worksystem, that is, by both human and computer.

This statement embodies the Dowell and Long (1989) distinction between the behavioural system, that is, the interactive worksystem, that performs work and the world of work, the domain of application, within which the work is performed. The distinction is shown in Figure 2.1. Pa is a function of the actual quality of the products of work within a particular domain of application (Qa) and the actual costs incurred by a particular worksystem (Ka).

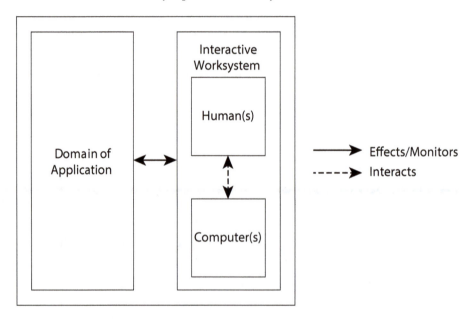

Figure 2.1: Behavioural system and work distinction (following Stork, 1999).

2.1.2 CONCEPTION OF SUBSTANTIVE HCI ENGINEERING DESIGN PRINCIPLES

HCI-EDPs express knowledge to support HCI design practice effectively in the provision of artefact specifications to satisfy user and any other requirements.

2.1 CONCEPTION OF SUBSTANTIVE HCI ENGINEERING DESIGN PRINCIPLES 11

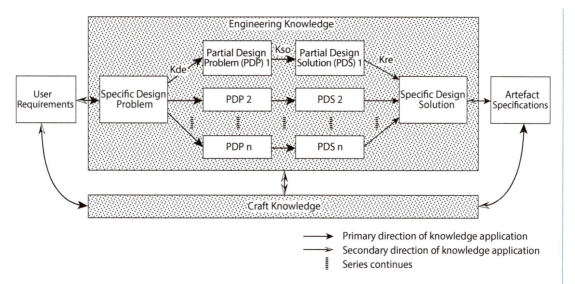

Figure 2.2: HCI engineering design practice (following Stork, 1999).

The HCI engineering design knowledge, applied during practice, is conceptualised as producing: a specific design problem operationalisation; partial design problem operationalisations; partial design solution operationalisations; and a specific design solution operationalisation, as shown in Figure 2.2. The partial design problem and solution operationalisations are the instantiations of a general design problem and its general design solution, as shown in Figure 2.3. The specific design problem and solution operationalisations represent the scoping of an HCI discipline as engineering. The partial design problem and solution operationalisations represent the application of HCI engineering design knowledge. They are "partial" because they solve only part of the specific design problem.

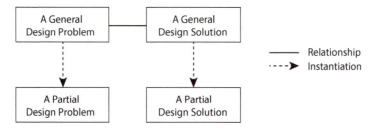

Figure 2.3: HCI engineering design knowledge (following Stork, 1999).

Dowell and Long (1989) distinguish between declarative (substantive) and procedural (methodological) HCI-EDPs. This case study restricts itself to declarative rather than to procedural HCI-EDPs. Procedural principles prescribe the methods for solving a general design problem optimally. Declarative principles prescribe the features and properties of artefacts, or systems that constitute an optimal solution to a general design problem.

Three types of HCI-EDPs are identified:

1. Decomposition knowledge (Kde) is conceptualised as the means of instantiating a partial design problem from a specific design problem. Kde requires declarative knowledge of the general design problem of which the partial design problem is the instance (see Figure 2.3).

2. Solution knowledge (Kso) is conceptualised as the means of instantiating a partial design solution from a partial design problem. Kso requires declarative knowledge of the general design solution to the general design problem identified in Kd (see Figure 2.3).

3. Recomposition knowledge (Kre) is conceptualised as the means of instantiating a specific design solution from partial design solutions. The assured prescription of the declarative Kso implies that recomposition would be prescribed, and so no declarative knowledge is required for Kre.

The specific design problem and solution may not be required for HCI engineering design practice. It may be possible to instantiate a partial design problem from the user or any other requirements using Kde. It may also be possible to instantiate part of the artefact specification using Kre. However, it is expected that the specific design problem and solution are required, at least, as the object of further research.

The declarative knowledge required for Kde, Kso, and Kre is a general design problem and its general design solution, conceptualised by a general desired performance and a general actual performance, respectively. A general design problem and its general design solution are general over types of user, types of computer, and types of domain of application. Desired performance and actual performance are conceptualised further below, following Dowell and Long (1989).

2.1.3 CONCEPTIONS OF THE GENERAL HCI DESIGN PROBLEM AND GENERAL HCI DESIGN SOLUTION

The general HCI design problem requires a statement of the desired performance for the desired worksystem. Whereas a statement of the general design solution requires a statement of the actual performance for the actual worksystem.

Desired performance and actual performance are conceptualised in the following sections. Important occurrences of the concepts are highlighted in bold for easy identification. The Dowell and Long (1989) concepts appear in italics. Quotations are from the latter source.

2.1 CONCEPTION OF SUBSTANTIVE HCI ENGINEERING DESIGN PRINCIPLES   13

2.1.3.1 Conception of Desired Performance

The **desired performance**, Pd, is conceptualised as a function of the **desired *quality*** of the products of work, Qd, within the domain of application and the acceptable or **desired *costs***, Kd, incurred by the worksystem.

The ***worksystem* boundary criteria** allow statement of the behavioural system, constituting the ***worksystem***. That system "whose purpose is to achieve and satisfy common goal[s]." The ***domain* boundary criteria** allow expression of the world of work that constitutes the domain of application. The latter is determined by the requirement to express the common goals.

2.1.3.2 Conception of Actual Performance

Actual **performance**, Pa, is conceptualised as a function of the **actual *quality*** of the products of work, Qa, within the given domain of application and the current or **actual *costs***, Ka, incurred by the worksystem.

The ***worksystem* boundary criteria** and ***domain* boundary criteria** are the same as for desired performance.

2.1.3.3 Conception of Desired Quality

Dowell and Long (1989) conceptualise the world of work as consisting of ***objects***, that have ***attributes*** having a set of possible ***states*** (defining their affordance for change). The ***desired quality*** of the products of work of the worksystem are conceptualised as transformations of states of attributes of objects that are desirable. The latter are termed ***product goals***. These objects and their attributes are conceptualised as ***abstract*** or ***physical***, and ***related*** or ***unrelated***. The transformations described by a product goal can be identified for each attribute, and these transformations are termed ***task goals***.

Dowell and Long describe abstract and physical attributes of objects. "Abstract attributes of objects are attributes of information and knowledge" and "physical attributes of objects are attributes of energy and matter." They also propose that "different attributes of an object emerge at different levels within a hierarchy of levels of complexity." In general, abstract attributes emerge at a higher level than physical attributes. Similarly, "objects are described at different levels of specification, commensurate with their levels of complexity." Furthermore, attributes of objects are related to attributes of other objects both between and within levels of complexity.

2.1.3.4 Conception of Actual Quality

The **actual *quality*** of the products of work, achieved by the worksystem, are conceptualised as for desired quality. The transformations of states of attributes of objects, that are achieved are termed **product *achieved* goals**. The transformations for each attribute are termed ***task* achieved *goals***.

14   2. INTRODUCTION TO INITIAL HCI ENGINEERING DESIGN PRINCIPLES

2.1.3.5 Conception of Desired Costs

Dowell and Long (1989) conceptualise the worksystem (the behavioural system) as "human and computer *behaviours* together performing work." Human behaviour is considered as purposeful and computer behaviour is considered as purposive. Human behaviours correspond with the transformation of objects in a domain and any expression of them must "at least be expressed at a level commensurate with the level of description of the transformation of objects in the domain." The conceptualisation applies to both computer and worksystem behaviours.

These behaviours can be *abstract* or *physical*. Abstract behaviours "are generally the acquisition, storage, and transformation of information. They represent and process information at least concerning: domain objects and their attributes; attribute relations and attribute states, and the transformations required by goals." Physical behaviours express abstract behaviours and are "related in a hierarchy of behaviour types."

Dowell and Long conceptualise the user as having *cognitive*, *conative*, and *affective* behaviours. "The cognitive aspects of the user are those of knowing, reasoning, and remembering. The conative aspects are those of acting, trying and persevering. The affective aspects are those of being patient, caring, and assuring."

Dowell and Long conceptualise humans and computers as "having (separable) *structures* that support their (separable) behaviours." Furthermore, "Human structures may be *physical* (neural, biomechanical, and physiological) or *mental* (representational schemes and processes)." Similarly, computer structures may be *physical* or *abstract*.

Dowell and Long (1989) claim that "work performed by worksystems incurs resource costs." They identify resource costs as behavioural or structural and associated with the human or the computer (separately). These costs can be further associated with abstract (mental) and physical behaviours or structures. Examples of resource costs related to the human are: physical workload for *human physical behavioural costs*; mental workload for *human abstract (mental) behavioural costs*; physical development and deterioration for *human physical structural costs*; and mental development and deterioration for *human abstract (mental) structural costs*. Examples of resource costs related to the computer are: energy emission and consumption for *computer physical behavioural costs*; software and functional resource (transaction and access resources); usage for *computer abstract behavioural costs*; system (hardware) development and degradation for *computer physical structural costs*; and software and functional development (and degradation) for *computer abstract structural costs*.

The **desired** *costs* are conceptualised as the necessary resource costs of the worksystem to achieve the desired task quality.

2.1.3.6 Conception of Actual Costs

The actual costs are conceptualised as the actual resource costs of the worksystem to achieve the actual quality.

2.1.4 CONCEPTIONS OF THE SPECIFIC HCI DESIGN PROBLEM AND THE SPECIFIC HCI DESIGN SOLUTION

The conceptions of the specific HCI design problem and solution are operationalised from the conceptions of the general HCI design problem and solution. The specific HCI design problem and solution are particular, by definition, to a scenario of HCI design.

The specific desired performance is conceptualised as a function of the desired quality of the products of work within a particular domain of application. Also, the desired costs, incurred by a particular worksystem.

The specific actual performance is conceptualised as a function of the actual quality of the products of work within a particular domain of application, as well as the actual costs incurred by a particular worksystem

2.1.5 CONCEPTION OF SUBSTANTIVE HCI ENGINEERING DESIGN PRINCIPLES REVISITED

HCI EDPs achieve, or exceed, prescribed performance on application (Pa = Pd). The conceptions of a general HCI design problem and its general HCI design solution can be combined to produce a single conception of a substantive HCI-EDP. Any expression of the domain, actual task quality and actual costs are not required for a general design solution (or for its partial design solution). They will be the same as those for its general design problem (or for its partial design problem). Therefore, the only component of the actual performance of a general design solution, that is not expressed by the desired performance in its general design problem, are those structures and behaviours of the worksystem required to achieve that desired performance. A substantive HCI-EDP is conceptualised then as the desired performance of a general design problem and the structures and behaviours of its general design solution.

2.2 STRATEGY FOR DEVELOPING HCI ENGINEERING DESIGN PRINCIPLES

This section proposes a strategy for developing HCI-EDPs and compares it with alternative strategies. The strategy selected is identified and then specified in detail. The case study aims to operationalise the strategy by developing examples of initial HCI-EDPs. That is, ones which to be final

16   2. INTRODUCTION TO INITIAL HCI ENGINEERING DESIGN PRINCIPLES

need to be complete and validated empirically. The rationale for the selection of the application domain of planning and control is presented.

2.2.1   STRATEGY DEVELOPMENT

One way of developing substantive HCI-EDPs is to identify general relationships between specific design problems and their solutions. These general relationships are putative, that is, they require validation. Here, they are termed "initial" as opposed to "final" HCI EDPs. The identification of general relationships between specific design problems and their solutions requires the operationalisation of specific HCI design problems and their solutions from the conceptions of specific HCI design problems and specific HCI design solutions. Testing is needed to validate such initial HCI-EDPs.

The case study reports the acquisition of initial HCI-EDPs by applying this strategy. The latter is also assessed. Two specific design problems and their solutions (Cycle 1 and Cycle 2) are operationalised, as the minimum able to support any claim of generality.

Assessment of the strategy comprises: acquisition (or not) of initial HCI-EDPs; assessment of the status of such acquired principles; and discussion of the strategy and conceptions, following the acquisition (or not) of initial HCI-EDPs.

A conception here is a set of concepts and their relations, which are abstractions over a class of objects, based on their common aspects. Conceptualisation is the process of generating a conception. Operationalisation is the process of instancing a conception to produce an operationalisation. An operationalisation of a conception is a set of less abstract concepts (related to the concepts in the conception) that ultimately reference observables in the world.

Dowell and Long (1989) claim that, for the acquisition of the knowledge to support HCI-EDPs, the operationalisation of the specific design problems and solutions needs to be explicit and formal. Formal, here, is understood as having defined rules of syntax and semantics. It is, thus, understandable by some people for some purpose. Formality requires the metrication of the operationalisation of the conceptions of the specific HCI design problem and solution. Metrication is defined as the process of instancing an operationalisation to its limit to produce metrics. Metrics quantify the less abstract concepts of the operationalisation in an observable relation with the world.

To operationalise specific design solutions to specific design problems, the following method is applied.

1.   Appropriate user requirements are selected for each of the two development cycles (see § 2.2.7).

2.2 STRATEGY FOR DEVELOPING HCI ENGINEERING DESIGN PRINCIPLES

2. An artefact specification is developed to satisfy the user requirements for each of the two development cycles, using "best-practice" (see § 3.4 for Cycle 1 and § 4.2 for Cycle 2).

3. The specific design problem and its solution are operationalised, based on the user requirements and its artefact specification for each of the two development cycles (see § 3.4 for Cycle 1 and § 4.2 for Cycle 2).

HCI-EDPs are conceptualised as the desired performance of a general design problem and the structures and behaviours of its general design solution. The strategy, however, is not to limit the operationalisations to these concepts, so as to provide:

1. a check that the specific design solution is a solution to the specific design problem (that is, by explicit representation of the actual performance to be compared with the desired performance);

2. the establishment of the relationships between the specific design solution structures and performance (a check for the solution); and

3. the availability of the research products for further work, some of which might aim for a lower prescribed performance.

The case study is restricted to the acquisition of cognitive HCI-EDPs. That is, excluding conative or affective principles. Cognitive processes and representations are relatively well defined , compared to the others. The latter constitute objects for possible future research.

However, reference also needs to be made to: additional concepts, current solutions, and "best-practice."

As concerns "additional concepts" to operationalise specific design problems and solutions, these are derived from a conception of human-computer systems costs. Human-computer costs are poorly conceptualised by specific design problems and solutions (for example, relative to task quality). Section 2.3 proposes such a conception.

As concerns "current solutions," it is easier to operationalise existing, installed, specific design solutions than to operationalise specific design problems. The operationalisation of the former provides a basis for the operationalisation of the latter. The desired performance of the specific design problem is likely to be similar to the actual performance of the current solution. Both the initial HCI-EDP development cycles operationalise the current solution before the operationalisation of the specific design problem and its solution (see § 3.1–3.2). The selection of re-design user requirements supports the operationalisation of a current solution.

Current design knowledge can be applied to user requirements, as part of best practice. The selection of tractable user requirements ensures that the differences between the operationalisation of the current solution and the specific design problem are minimal. Also, that a specific design solution exists.

"Best practice" for developing design solutions is considered to be the best design practice at the time of HCI-EDP acquisition. At the time of the present case study, the latter was considered to include the application of a structured analysis and design method (MUSE - Method for USability Engineering; Lim and Long, 1994), the use of design guidelines and evaluations. However, there was no consensus, concerning HCI best practice at the time and so alternative design knowledge and practices could have been applied. Likewise, today's best practice would be expected to include design knowledge and practices, developed since the time of the case study. Such design knowledge and practices would be expected to include among others those propagated by Ritter et al. (2014), Hartson and Pyla (2018), Kim (2020), and Zagalo (2020). The key point, however, is not the particular best practice espoused per se, but its ability to support the particular solution of particular design problems, as required by the acquisition of HCI-EDPs.

2.2.2 COMPARISON WITH ALTERNATIVE STRATEGIES

The strategy for developing HCI engineering principles is characterised as "bottom-up" and cautious, or even sceptical (Stork, 1999). The cautious approach entails either that steady progress is made towards HCI-EDPs or that the research direction is abandoned. The cautious approach also accepts that HCI-EDPs may not be acquired initially.

An alternative "top-down" strategy would be to postulate operationalisable and testable HCI-EDPs, based on their conceptualisation. Such principles could then be operationalised and tested. The strategy is considered bold, given the current modest understanding of HCI-EDPs. The likelihood of identifying such a principle might be low. However, the effort for each attempt would be less than with the bottom-up strategy. The top-down strategy is rejected here. The low likelihood of identifying an HCI-EDP is insufficient to merit the effort of each attempt.

A "middle-out" strategy would be to develop a conception of the general design problem and solution for a simple design world, for example, of simple shapes, associated with a small set of requirements. However, scaling up would require the adoption of one of the other two strategies.

2.2.3 SCOPING THE RESEARCH USING THE POTENTIAL FOR PLANNING AND CONTROL HCI ENGINEERING DESIGN PRINCIPLES

The case study addresses the potential for planning and control HCI-EDPs. The latter rests on computer support for the following fields: military planning; aircraft flight planning and control; office administration; project management; business decision-making; and clinical decision-making. The latter may be associated with different HCI-EDPs. However, it is likely that, together, they offer promise for general HCI-EDPs for planning and control. Hence, the case study scope of the operationalisation of planning and control concepts (see § 3.2) for specific design problems and their solutions.

2.2.4 ACQUIRING POTENTIAL GUARANTEE

Ultimately, the guarantee or reliability of HCI-EDPs rests on the effectiveness of their support for design practice. Their validation is by testing. However, initial HCI-EDPs need to acquire potential guarantee or reliability to support validation. Following Dowell and Long (1989), effective HCI knowledge is conceptualised, operationalised tested, and generalised. Potential guarantee prerequisites are that:

1. initial HCI-EDPs be conceptualised, according to a conception of the general design problem of an engineering discipline of HCI;

2. such principles be operationalisations of those conceptions. The cycle operationalisations identify the concepts operationalised to support informal checking;

3. such principles be generalised. The generalisation here is over the two development cycles; and

4. such principles be tested. The testing here is the informal evaluation of the two development cycles.

2.2.5 SHORTER-TERM RESEARCH BENEFITS

The research strategy is to develop HCI knowledge for HCI practice in the longer term, as HCI-EDPs. Shorter-term research benefits may also accrue. For example, a medium-term research (and practice) benefit is a version of MUSE (Lim and Long, 1994) that supports a more complete, coherent, and consistent specification of the design problem and solution.

2.2.6 OVERVIEW OF MUSE

The overview is only sufficiently detailed for readers to understand its application in the case study. MUSE is a structured analysis and design method. It aims to improve HCI practice by providing support for the integration of human factors with existing structured methods for software engineering, such as JSD, Yourdon, or SSADM (Structured Systems Analysis and Design Method). The output of MUSE is the specification of an interactive system artefact. The software engineering method produces the specification of an implementable artefact, which incorporates the interaction artefact.

20 2. INTRODUCTION TO INITIAL HCI ENGINEERING DESIGN PRINCIPLES

MUSE supports design in a top-down manner, based on information derived bottom-up. Application progresses from the specification of general features of the tasks to be performed by the user, derived from analysis of the user requirements and from existing systems, to the specification of the details of the interaction artefact.

Figure 2.4 shows MUSE. It has three phases. The Information Elicitation and Analysis Phase supports the assessment and re-use of components of existing systems and the maintenance of the consistency of the design with the user requirements. The Design Synthesis Phase supports the conceptual design of the interaction artefact and the maintenance of the consistency of the design. The Design Specification Phase supports the detailed design of the interaction artefact. Checking and information exchange with the software engineering method occurs to ensure artefact implementability and consistency.

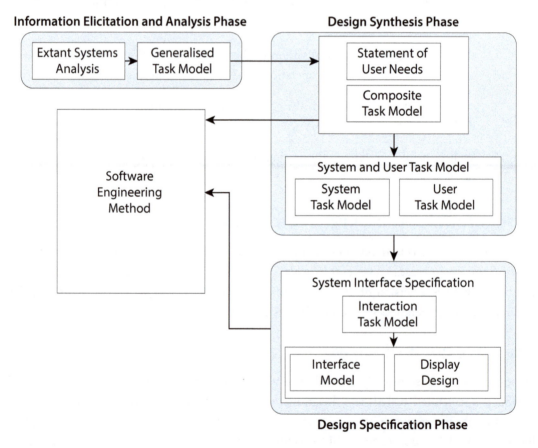

Figure 2.4: MUSE overview (following Lim and Long, 1994, cited by Stork, 1999).

2.2.7 DEVELOPMENT CYCLE USER REQUIREMENTS SELECTION CRITERIA

For the development cycles, tractable, re-design, and relatively simple user requirements are selected. The latter are more easily satisfied and so are more likely to yield initial HCI-EDPs. Nevertheless, the user requirements offered the development cycles the potential for repetition, access, interest, and generalisation. A questionnaire supported identification of potential user requirements. No other requirements were included, again to keep the scenario simple.

The user requirements are intended to be operationalisable, according to the conception of the specific design problem, to avoid concentration on the relationship per se between the user requirements and the operationalisation of the specific design problem. The rationale is as follows.

1. As concerns tractable and re-design user requirements, they are part of the development strategy (see § 2.2.1).

2. Concerning relatively simple user requirements, the conceptions of the specific design problem and solution are difficult to operationalise. So, relatively simple user requirements support relatively simple operationalisations. For Cycle 2, however, some increase in complexity over Cycle 1 is appropriate.

3. As concerns repetition, the user requirements are such as to maintain the design scenario and operationalisation.

4. As concerns access to the particular design (before, during, and after the design process), it should be easy to permit: cycle selection; best-practice design; evaluation, and the operationalisations.

2.2.7.1 Cycle 1 Selection

Cycle 1 Selection comprises user requirements and their comparison against criteria.

The user requirements are as follows.

> X, the home co-owner, expresses broad user requirements for home heating. X lives with Z, the second home co-owner. Their home heating problems are skewed towards being too cold, since observations occurred in winter. The user requirements follow.

> 1. Different tasks require different conditions for comfort. Any particular room is not always comfortable. The task being performed can change faster than the conditions in the room. This problem is more noticeable in the downstairs rooms, which cannot be controlled as individually as the others.

2. The two occupants of the house, X and Z, do not always require the same level of comfort.

3. The heating in the morning at weekends is nearly always too hot, while the occupants are still in bed and then too cold on rising.

4. If the occupants are up late or friends visit, then the house can become cold, unless the heating is switched back on.

5. If the occupants are out late, then the house is cold on their return.

6. If either or both of the occupants return by bicycle, they are usually too hot in the house. Occasionally, Z is too cold on their return by bicycle.

7. If X leaves after 8 a.m. or stays at home to work, then the house is too cold, until the heating is turned back on. If X and Z expect to be at home for a short time, then they often use the heating boost facility, which can result in them being too cold, if they are at home for longer than expected.

The current gas bill is acceptable for the resulting comfort. An increase could be considered acceptable for greater comfort. A decrease in the gas bill for the same comfort or better would be desirable.

As concerns the comparison of user requirements against the criteria, all are tractable re-design requirements with repetition, access, interest, and generalisation potential (see § 2.2.7).

The last identified user requirements are selected for Cycle 1, since they are relatively simple (being based primarily on X). They have repetition, appearing to be time-invariant (if taken for days of a specific outside condition or worse) with a reasonably constrained set of factors that are not invariant.

2.2.7.2 Cycle 2 Selection

The Cycle 2 Selection criteria are as follows for three possible types of scenario.

1. X's home. One of the user requirements not selected in Cycle 1 could be selected for Cycle 2. However, they were rejected for Cycle 1 selection.

2. X's car. It was felt that for generalisation with Cycle 1, home would be more appropriate.

3. Another home, whose occupants are known to the researcher, which would be similar but different, together improving the potential generality.

Cycle 2 User Requirements are selected from the third option, those in another home, whose occupants are known to the researcher.

The following broad user requirements were identified after discussion with the occupants of the home, *P* and *V*. These problems are skewed towards being too cold, since most of the observation was performed in winter, matching Cycle 1. User requirements follow.

1. The main house is too cold, if only one of its boilers is started for the early mornings, since the following areas are always accessed in the morning and their radiators are supplied by different boilers: the kitchen; the front porch (for the mail and newspapers); and the downstairs toilet.

2. Study 1 and Study 2 (studio) are always cold for sedentary working, since the radiators are badly located for rooms with external walls. The rooms are comfortable, once warmed using fan heaters.

3. The sitting room can be cold on winter evenings, particularly if the boiler supplying the radiators in the sitting room has been off during the day.

4. The dining room can be too hot, when there are many people in it. It is undesirable to open the window, since it faces the prevailing wind.

5. The kitchen is a comfortable room with thick walls that retain the heat. However, it can get too hot during cooking, particularly in the summer, but also in winter. The windows are all fixed with security locks.

6. *P* can feel cold, while working, as they require a warmer temperature to work than when they perform other, more physical, tasks (for example, cooking or house repairs), and warmer than *V* requires.

7. *P* often works in the cottage, a small property attached to the main house, since they can control the heating more easily. It is separate from *V*'s heating requirements. *P* finds the controls easier to use. *P* usually knows in advance that they will be working in the cottage. *P* has to walk across the garden to turn the heating on, or up, before returning to work after the cottage has warmed up. *P* normally leaves the heating on in the main building for their return (even if *V* is out).

8. *V* tends to turn the heating off, if they are going out for the day or longer. *P* tends to leave it on, so that it is warm on their return.

9. *V* turns the heating off on April 1st for summer. *P* would prefer it on, since they are sometimes cold in summer.

10. The timers are all difficult to adjust, being mechanical, situated separately, and in dark corners (one of them is in a cupboard). The occupants feel that the controls

require moving and improving (with separate weekend times and digital controls). They have installed the wiring to put the two main house controllers in the lobby.

11. Ventilation is very poor throughout the house.

The heating costs seem high, but there are no standards for comparison. Any reduction would be welcome. Any improvement should not cost more than the gas bill reduction and any increase in house value.

As concerns the comparison of user requirements against the criteria, the following are selected. The kitchen can get too hot during cooking particularly in the summer, but also in winter. The kitchen is a comfortable room with thick walls that retain the heat. The windows are all fixed with security locks. These tractable re-design user requirements have good access, interest, and generalisation potential. They are relatively simple, since there are few conflicting needs and they are not based mainly on the technology. They are marginally more complex than the Cycle 1 User Requirements.

2.3 CONCEPTION OF HUMAN-COMPUTER SYSTEMS

Section 2.2 proposes a strategy for developing HCI-EDPs. Part thereof is the operationalisation of specific design problems and their solutions from their associated conceptions. Since human-computer system costs are poorly conceptualised, relative to task quality in those conceptions, an initial conception of human-computer systems and their costs is proposed.

2.3.1 INTERACTIVE WORKSYSTEM COSTS

According to Dowell and Long (1989), costs appertain to human or computer and are separable. That is, the human and computer costs are conceptualised individually. However, they can also be integrated.

To enable the separable costs to be integrated, the integrated worksystem costs are further conceptualised before their separate conception. Human and computer behavioural costs are conceptualised from derivation of each behaviour occurrence. Dowell and Long conceptualise human and computer structural costs as initial and ongoing. The latter arise from the initial processes and representations present, but required at the start of the design problem or solution. Initial processes are conceptualised as including the ordering of the behaviours, during the design problem or solution. Ongoing structural costs arise from the development or change in state of processes and representations, during the design problem or solution.

All costs are initially conceptualised as unitary (Dowell, 1998) and, so, non-dimensional. Each behaviour occurrence incurs one unit cost. Each initial process and representational structure incurs one unit cost. Each ongoing structural change incurs one unit cost. Non-unitary costs are also possible.

2.3.2 POTENTIAL HUMAN COGNITIVE STRUCTURES

Long and Timmer (2001) and Timmer and Long (2002) propose an "operator mental architecture," based on a computational cognitive architecture. The latter is selected here because the cognitive architecture is relatively simple. The process of "problem solving" is not conceptualised further. It employs some concepts from the general design problem of HCI (Dowell and Long, 1989), including the distinction between domain and worksystem, user ("operator") and computer ("device"), and structure and behaviour. It has been employed for design diagnosis in air traffic management. Timmer and Long describe the architecture as follows: "The ... architecture 'distinguishes four classes of mental structure: storage; process; transducer, and representational. ... Three major storage structures are specified: long-term memory; working memory; and a goal store, accommodating a single active goal. Eleven process structures are loosely associated with particular storage structures: 'decay' and 'store' in long-term memory; 'form', 'pop', 'suspend,' and 'reactivate', for goal management in the goal store; and higher level processes of 'categorise', 'problem-solve,' and 'evaluate' in working memory. A single mental processor is assumed in working memory. An input transducer, with an associated 'encode' process, maps environmental stimuli into a mental code. An output transducer, with an 'execution' process, maps an action specification into physical behaviour."

Figure 2.5 shows the cognitive architecture and its relationship with the human physical architecture, which is described next.

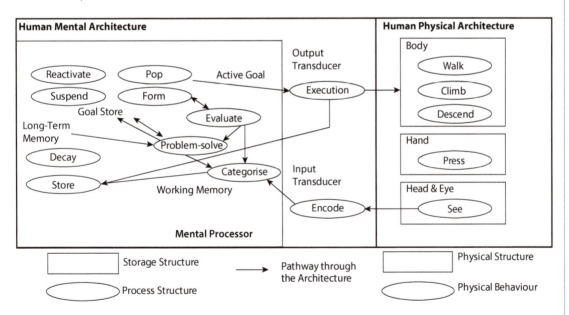

Figure 2.5: Human architecture (following Timmer 1999, cited by Stork, 1999).

26   2. INTRODUCTION TO INITIAL HCI ENGINEERING DESIGN PRINCIPLES

2.3.3   POTENTIAL HUMAN PHYSICAL STRUCTURES

The emphasis here is on cognitive structures and behaviours. The human physical architecture is conceptualised as any part of the human body, or the body itself, required for operationalising the specific design problem and solution. Figure 2.5 shows the human architecture for development Cycle 1.

2.3.4   POTENTIAL COMPUTER ABSTRACT STRUCTURES

A computer architecture is conceptualised in a similar manner to the human architecture. Figure 2.6 shows the computer (Von Neumann-based) architecture.

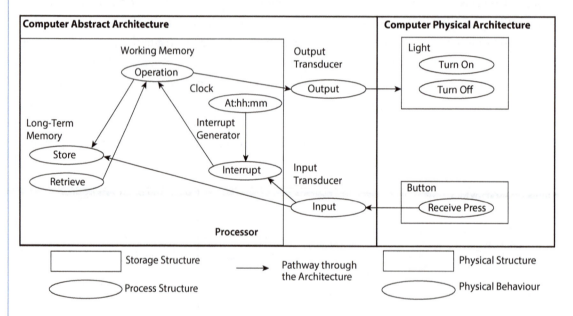

Figure 2.6: Computer architecture (following Stork, 1999).

2.3.5   POTENTIAL COMPUTER PHYSICAL STRUCTURES

The computer physical architecture is conceptualised as any device/artefact required for operationalising the specific design problem and solution. Figure 2.6 shows the computer architecture for Cycle 1.

The unitary costs of the human-computer system arise from the occurrence of each behaviour, each initial structure and each ongoing structural change.

The potential human-computer system structures/architectures are conceptualised. The structures support the potential human-computer system behaviours. The structures and behaviours are considered potential, because they offer an initial view to be validated by HCI EDPs.

REVIEW

The chapter presents a conception of declarative HCI-EDPs, together with an associated instance-first strategy for their development. The latter is scoped by the potential for planning and control HCI-EDPs. The strategy requires cycles of current HCI best-practice development and operationalisation of specific design problems and solutions. Two development cycles are proposed. The potential human-computer system structures, or architectures, are conceptualised.

2.4 PRACTICE ASSIGNMENT

2.4.1 GENERAL

Read § 2.1, concerning the conceptions of declarative HCI-EDPs.

- Check the conceptions informally for completeness and coherence, as required by the case study of domestic energy planning and control.

- The aim of the assignment is for you to become sufficiently familiar with the conceptions to apply them subsequently and as appropriate to a different domain of application, as in Practice Scenarios 2.1–3.

Hints and Tips

Difficult to get started?

Re-read the assignment task carefully.

- Make written notes and, in particular, list the conceptions, while re-reading § 2.1.

- Think about how the conceptions might be applied to describe a novel domain of application.

- Re-attempt the assignment.

Test

List from memory as many of the conceptions as you can.

Read § 2.2, concerning the strategy for developing HCI-EDPs.

28 2. INTRODUCTION TO INITIAL HCI ENGINEERING DESIGN PRINCIPLES

- Check the sections informally for completeness and coherence, as required by the case study of domestic energy planning and control and complete as for Read § 2.1 earlier.

Read § 2.3, concerning the conception of human-computer systems.

- Check the sections informally for completeness and coherence, as required by the case study of domestic energy planning and control and complete as for Read § 2.1–2 earlier.

2.4.2 PRACTICE SCENARIOS

Practice Scenario 2.1: Applying Conceptions for HCI Engineering Design Principles to an Additional Domain of Application

Select a domain of application, with which you are familiar or which is of interest to you or preferably both. The domain should be other than that of domestic energy planning and control.

- Apply the conceptions for the domestic energy planning and control (see § 2.1) to the novel domain of application. The description can only be of the most general kind— that is at the level of a conception. However, even consideration at this high level can orient the researcher towards application of the conceptions to novel domains of application. The latter are as might be required subsequently by their work. The research design scenario is intended to bridge this gap.

Practice Scenario 2.2: Applying a Strategy for Developing HCI Engineering Design Principles to an Additional Domain of Application

Select the same novel domain of application as for Practice Scenario 2.1 and complete as for the previous section.

Practice Scenario 2.3: Applying a Conception of Human-Computer Systems to an Additional Domain of Application

Select the same novel domain of application as for Practice Scenarios 2.1–2 and complete as for the two previous sections.

CHAPTER 3

Cycle 1 Development of Initial HCI Engineering Design Principles for Domestic Energy Planning and Control

SUMMARY

This chapter introduces the development of initial HCI-EDPs for the application domain of domestic energy planning and control. The chapter comprises: the operationalisation of specific design problems and solutions; the conception of planning and control; Cycle 1 best-practice development; and Cycle 1 operationalisation.

3.1 OPERATIONALISING SPECIFIC DESIGN PROBLEMS AND SOLUTIONS

Section 2.1 presents conceptions to support operationalisations of specific design problems and solutions. Section 2.3 presents a conception of human-computer systems and their costs. This section presents frameworks for operationalisations of these conceptions. The frameworks include the layout and scope of diagrams and tables for the metrication of the operationalisations.

Frameworks are presented for task quality, including the domain, and for worksystem costs, and the associated human-computer structures and behaviours. Composite structures are defined as groups of processes, which occur repeatedly.

3.1.1 FRAMEWORK FOR TASK QUALITY

The states of the task quality, product goals and task goals are conceptualised by numerical or Boolean values over time. The relationships (between and within the hierarchy of complexity) are conceptualised by formulae.

The domain concepts from the Dowell and Long (1989) conception of the HCI general design problem are considered sufficient for operationalisation. The domain diagram represents the objects having abstract attributes, relationship and physical attributes. Figure 3.1 shows the key for domain diagrams.

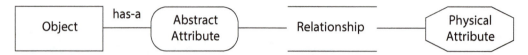

Figure 3.1: Domain diagram key (following Stork, 1999).

The task goals, product goals, and task quality Boolean values are documented as attributes of the objects, together with their relationship with the other attributes. The relationships are intended to be mathematical, including the Boolean logical operations.

The states of the attributes for each instance of the current or actual design are recorded in a "state stream table." Table 3.1 shows the headings to be used for the state stream table, with example entries.

Table 3.1: Domain State Stream Table Key

Time	Event	Attribute 1	Attribute 2	Attribute 3	Attribute 4	...	Task Goal 1	Task Goal 2	Product Goal 1	Task Quality
0:00	1	3.4	10°	TRUE	-62	...	TRUE	FALSE	FALSE	FALSE
...	2									

The time column shows a progression of the design problem or solution. The time interval selected is dependent on the rate of change of the domain and worksystem. The event column shows the ordering of domain changes, including their occurrence within the time frame. It is required later for the framework for worksystem costs. The first row shows the initial states. The state stream table supports better specification of the relationships between the attributes during the design problem or solution. Further, it can be used to identify a formula for the state over time. If such a formula is identified, it can be used to calculate the state over time.

Operationalising the task quality of a worksystem could be an attempt to operationalise a very general purpose, such as a human's existence. However, for the present operationalisations, only more specific purposes, such as comfort, need to be operationalised. This constitutes a boundary meta-assumption.

3.1.2 FRAMEWORK FOR INTERACTIVE WORKSYSTEM COSTS

The diagram for the worksystem shows the process structures (which support the behaviours) using a MUSE-like notation (Lim and Long, 1994) and the representational structures using the domain key. The potential behaviours, supported by the process structures, that change the states of the domain are linked with a line to the domain state that can be changed in the domain diagram.

The worksystem effects state changes over time. It is often difficult to identify the time between behaviours, for example, to identify the time between seeing an object and its categorisation. The concept of "events" is introduced to operationalise the ordering of behaviours, without distinguishing the time. A new event occurs on every behaviour and the time recorded against that behaviour.

The structures of the occurring behaviours and their associated costs are placed as the headings in a table to match that of the domain table above. This "structure and behaviour streams" and costs table is shown in Table 3.2 with example entries. For the structures, the change in the state are marked against time and event. For the behaviours, the occurrence of the behaviour are marked against time and event. The cost contribution of the structures and behaviours are shown in the first rows of the table. This cost contribution is the abstract and physical costs of the structure state change or behaviour occurrence (and development), separated into abstract and physical. The behaviour occurrences, structure state changes, and domain state changes can be related by formulae. The costs columns can then be calculated by formulae.

Table 3.2: Structure and behaviour streams and costs table key

Time	Event	Abstract beh.s		Physical beh.s		Costs		Abstrac struct.s	Struct. 2	etc.
		Beh. 1	Beh. 2	Beh. 1	Beh. 2	Cost 1	Cost 2	Struct. 1		
		Cost contrib. for Beh. 1	Cost contrib. for Beh. 2	Cost contrib. for Beh. 1	Cost contrib. for Beh. 2			Cost contrib. for Struct. 1	Cost contrib. for Struct. 2	...
0:00	1	FALSE	TRUE	FALSE	TRUE	1	3	0	TRUE	...
...	...	...	...	...	...	...	...	...	...	...

The time interval to select is dependent on the rate of change of the worksystem as well as that of the domain.

3.1.3 COMPOSITE STRUCTURES

Composite structures are conceptualised as groups of processes that occur repeatedly in the same operationalisation or across operationalisations. They may be used as process structures, in place of the repeated process structures. Composite structures can be given parameters.

Composite structures reduce the size of the diagrams and tables for the operationalisations, so improving their readability and development. They represent low-level structural generality within and between the operationalisations.

Table 3.3 shows the composite structures that are developed for the operationalisations. The planning and control composite structures (H:StMon, H:StSubPlan, etc.) refer to planning representations (CDc, CDd, CWd, etc.), which are conceptualised in the next section.

Table 3.3: Composite structures	
Composite Structure	**Description**
H:FP:X	Human forms goal, other behaviours occur, then goal is popped.
H:FS:X	Human forms goal, other behaviours occur, then goal is suspended.
H:RS:X	Human resumes goal, other behaviours occur, then goal is suspended.
H:RP:X	Human resumes goal, other behaviours occur, then goal is popped.
H:FxP:X	Human forms a goal to encode or execute X, encodes or executes X, then pops the goal.
C:IISO:X	Computer inputs X.
C:O:X	Computer outputs X.
H+C:Change gas:X, Change	Human and computer (cooker) change the gas of X (a ring or the oven) by change amount.
H:StMon:X,Y	Human collects information through sight, updates the CDc planning representation, and decides whether to change the plan.
H:StSubPlan:X,Y	Human updates the CDd planning representation, and then updates the CWd planning representation.
H:StMonA:W,X,Y,Z	Human collects information, updates the CDc planning representation, and decides whether to change the plan.
H:StMonB	Human updates the CWc planning representation.
H:StSubPlanA:W,X,Y	Human updates the CDd planning representation.
H:StSubPlanB	Human updates the CWd planning representation.
H:ShSubPlan:W	Human updates the CWd planning representation by writing.
H:StShSubPlan:W,X,Y	Human updates the CWd planning representation by either writing or mental storage.

3.2 CONCEPTION OF PLANNING AND CONTROL

Section 2.2 presents the rationale for scoping the case study to planning and control HCI-EDPs. This section presents an initial conception of planning and control. The latter supports the operationalisation of planning and control for the specific design problems and their solutions.

Conceptions of planning and control are to be found in the HCI, psychology, and artificial intelligence (AI) literature. The present conception accords priority to conceptions claiming design guidance.

3.2.1 CONCEPTIONS OF PLANNING AND CONTROL CLAIMING DESIGN GUIDANCE

The research claiming design guidance divides into that which identifies plans as being in the domain and which identifies plans as being representations in the worksystem. Planning is identified as occurring in the worksystem. Control is identified as either occurring in a different worksystem from the planning worksystem or occurring in the same worksystem that has performed the planning.

3.2.1.1 Plans in the Domain

Colbert (1994) proposes a design for a menu structure for planning systems. Rules systematically relate the menus to the planning. The general menu structure is instantiated by rules for two types of planning—the off-loading of men and equipment during amphibious operations and the planning of attacks with surface-to-surface guided weapons. The instantiated menus are evaluated, leading to revisions of the rules and the general menu design. The design guidance is explicit, as rules.

3.2.1.2 Plans in the Interactive Worksystem

Dowell (1993) develops design guidance based on the description of a planning and control worksystem that manages air traffic. The worksystem comprises the cognitive representations and processes that support planning and control behaviours. The representations include the current and future state of the air traffic management domain. The processes are developed on the basis of artificial intelligence planning.

Dowell separates planning processes from control processes. Cognitive representations express such plans. The latter contain states of the aircraft: the current, projected, planned, and goal states, plus planned interventions for the aircraft. The aircraft state is represented as the attributes of the domain. Dowell identifies the latter as the position, altitude, speed, and heading of the aircraft at a particular time (PASHT). The current state of the aircraft is its PASHT value, including that on leaving the sector, if no interventions are made. The planned states of the aircraft are the expected PASHT values given the planned interventions. The goal state of the aircraft is a desired

PASHT value on leaving the sector. The interventions are representations of the processes that the worksystem intends to execute to achieve the planned states.

Dowell claims air traffic management is a dynamic domain, having processes, which change state over time even without controller/pilot intervention. This reduces the time available for planning. Following artificial intelligence, Dowell characterises the worksystem as: a reactive planner, interleaving planning and control; a hierarchical planner, planning at a higher level than basic cognitive processes; and a nonlinear planner, which does not necessarily represent the planned processes linearly.

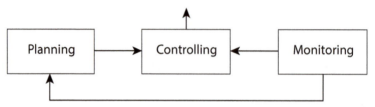

Figure 3.2: Abstract description of the planning and control behaviours of an interleaved planner (following Linney, 1991, cited by Stork, 1999).

Dowell follows Linney (1991) in identifying the interleaved planning, control and monitoring processes. These are shown in Figure 3.2. Dowell identifies processes for the worksystem that are less abstract, and relates those processes to the representations. Table 3.4 shows these processes and their related representation changes.

Table 3.4: Planning and control processes of an air traffic management worksystem (following Dowell, 1993, cited by Stork, 1999)

Process	Representation
Monitoring Behaviours	
generate	current airtraffic event (PASHT attribute values)
generate	current vector (actual and projected task attribute values)
generate	goal vector (goal task attribute values)
evaluate	current vector
Planning Behaviours	
generate	planned vector (planned task attribute values)
evaluate	planned vector
generate	planned interventions (PASHT attribute values)
Controlling Behaviour	
generate	execution of planned intervention (issue instruction)

The following design guidance is the result (Dowell, 1993).

1. The planned vector of a plane is not evaluated exhaustively with respect to safety. Improving its evaluation would improve the performance. The designer should highlight those aircraft with proximal projected vectors or train the controller in conflict search procedures.

2. The rate of plan moves is slow. Improving the rate would improve performance. The designer should ensure that the mental representation of the planned vector and the paper representation be closer.

3. The construction of current and goal vectors is acceptable. Change would reduce performance. The designer should ensure that the current flight strip spatial organisation of aircraft is retained.

4. The evaluation of planned vectors with respect to safety is adequate. Change would reduce performance. The designer should ensure that the flight information for proximal aircraft within the controller's planning horizon should be displayed together, as with the current flight strips.

3.2.2 CONCEPTIONS OF PLANNING AND CONTROL WITH NO CLAIMS FOR DESIGN GUIDANCE

Conceptions for which there is no explicit claim for design guidance follow. These conceptions may underlie design guidance research. Also, the operationalisation of planning and control may require their associated concepts.

As concerns HCI, Norman (1993) states that "for many everyday tasks, goals and intentions are not well specified: they are opportunistic rather than planned." A plan includes specifying goals and intentions, where intentions appear to lead to action.

As concerns town planning, Friend and Jessop (1969) identify planning as being "required for non-trivial action decisions, that is, prior elaboration of potential actions is required for them to be assessed." They also note that "it is [in public planning] exceptionally difficult to formulate strategies in advance, which are sufficient to cope with all conceivable contingencies … in these circumstances, planning must become in some degree an adaptive process."

As concerns psychology and AI, the Hayes-Roth et al. (1988) model of planning claims to be "computationally feasible and psychologically reasonable." Planning is "the process by which a person or a computer program formulates an intended course of action." Planners may make decisions about the contents of the plan in different ways. They may make abstract decisions about the "gross features of the plan" to guide decisions about the details or vice-versa.

The Hayes-Roth et al. planning model contains independent and asynchronous "specialists." The latter propose decisions for a tentative plan. The latter appears on a "blackboard" by which the specialists communicate. The plan indicates the "actions the planner actually intends to take." It stops being tentative, when the planner "accepts" the overall plan. This is after "plan evaluation, the analysis of likely consequences of hypothesised actions." Presumably, acceptance occurs when the plan evaluation passes some threshold. A "meta-plan" orders the execution of the specialists, a process of "situation assessment, analysis of the current state of affairs" and plan evaluation.

As concerns AI, Alterman (1988) describes adaptive planning instantiated in a system called PLEXUS—"the problem of adaptive planning … is to take a prestored plan … and apply it to a novel set of circumstances."

3.2.3   INITIAL CONCEPTION OF PLANNING AND CONTROL

The conceptions outlined above vary in their explicitness, completeness, coherence, their operationalisation in design guidance, and the claims for the latter.

The conception developed here aims:

1.   to be inclusive of the other conceptions of planning and control to ensure the widest potential to develop planning and control HCI-EDPs. For example, AI conceptions offer the best potential for operationalising computing planning and control;

2.   to decide between alternatives by selecting those with the better operationalised design guidance and has stronger claims; and

3.   to relate the planning and control conception to the conception of the general design problem.

The initial conception follows. Descriptions of planning and control are divided into those, which consider planning and control to be separate worksystems (for example, Colbert, 1994) and those which do not (Dowell, 1993). The distinction should be concerned with: the scope of the system to be designed; the knowledge to be acquired; or both. However, it appears to have an additional relationship with the "planning horizon," the length of time available before control must be performed, and, perhaps, therefore, with the design guidance. The conception offered here attempts to relate the two aspects.

Colbert's conception is an example of the separation of planning and control into separate worksystems. In the first, the Domain of Military Plans is separated from the Worksystem of Military Planning. In the second, the Domain of Armed Conflict is separated from the Worksystem of Armed Conflict.

Colbert fails to identify the relationship between the two worksystems. There are several alternatives (shown in Figure 3.3). Figure 3.3a shows plans specifying the desired states of the control

3.2 CONCEPTION OF PLANNING AND CONTROL

domain. Figure 3.3b shows plans specifying the behaviours of the control worksystem. Figure 3.3c shows plans specifying the (perhaps initial) contents of representations—of the desired states of the control domain and of the planned behaviours of the control worksystem being designed.

The first alternative is an analysis that would probably need to be performed during design of the control worksystem. The second is part of the specification to be produced during design of the control worksystem. The third could similarly be part of the specification to be produced during design of the control system or could represent a logical separation of planning and control for the application (and potentially therefore, the acquisition) of design knowledge.

Colbert's work does not suggest the latter. The former can be understood as the same as Figure 3.4a, if the plan representation content is understood to be abstract structures. The latter approach is adopted here. Colbert does not state which of the alternatives is intended, so Figure 3.3 contains Colbert's diagram representing these three alternatives. Figure 3.4 shows a generalisation of Figure 3.3. Figure 3.5 is termed "desired states and structures planning and control" (DSSP&C).

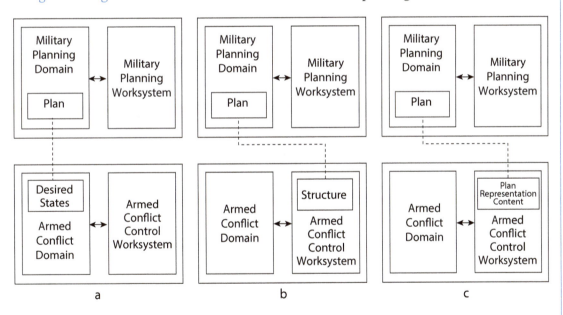

Figure 3.3: a, b, and c. Alternative representations of Colbert's planning and control (following Colbert, 1994, cited by Stork, 1999).

38 3. CYCLE 1 DEVELOPMENT OF INITIAL HCI ENGINEERING DESIGN PRINCIPLES

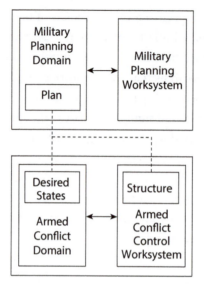

Figure 3.4: Composite representation planning and control (following Colbert, 1994, cited by Stork, 1999).

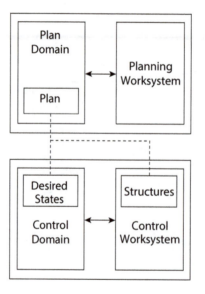

Figure 3.5: Desired states and structures planning and control (DSSP&C) (following Stork, 1999).

Dowell takes a different approach; see Figure 3.6. The approach is compatible and Figure 3.7 shows a general version of the overall target of planning and control as the control work. This general version can be represented, albeit in a more decomposed manner, by DSSP&C; see

Figure 3.8. Therefore, DSSP&C is taken as the basis for the planning and control conception here. The target concept is generalisable, as demonstrated by Stork et al. (1998), who apply it to training and emergency management.

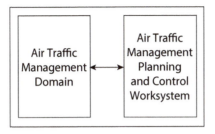

Figure 3.6: Planning and control (following Dowell, 1998, cited by Stork, 1999).

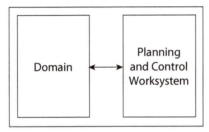

Figure 3.7: Generalised planning and control (following Stork, 1999).

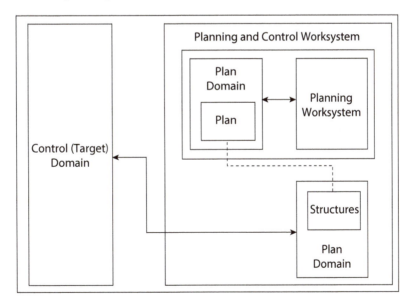

Figure 3.8: Planning and control represented in terms of DSSP&C (following Dowell, 1998, cited by Stork, 1999).

3.2.3.1 Control Domain and Control Worksystem

The control domain is conceptualised as for the domain in the specific design problem and solution conceptions. Thus, the control domain contains desired states for a desired performance operationalisation and actual states, for actual performance operationalisation

The control worksystem is conceptualised as for the worksystem in the specific design problem and solution conceptions. Thus, the control worksystem contains structures.

3.2.3.2 Plan Domain

Colbert (1994) identifies plan and sub-plan objects in the domain of plans for armed-conflict. Plan objects are "a representation of the goal states of [control] domain objects and/or desired future behaviours of a control worksystem." Sub-plan objects are "a specification of lower level goal states of [control] domain objects and/or desired future behaviours of a control worksystem."

Colbert further states that plans and sub-plans have attributes of scope, view, and content types. The scope types are: time_scope, "the period of time to which content applies"; object_scope, "the [control] domain objects to which content applies"; and behaviour_scope, the "control worksystem behaviours to which content applies." The view types are: view_type, "the type of representation"; view_content_options, "selections of content to be expressed in a representation"; and view_format_options, "variations in the physical representation of content." The content types are: content, "the specification of goal states of … [control] objects, and/or the behaviour of … control worksystems." The states of the attributes support the representations of the plan and sub-plan objects.

Colbert's conception is adopted here. The content is redefined as "the specification of desired states of control objects and/or the behaviours of control worksystems."

3.2.3.3 Planning Worksystem

The primary representation of the plan requires the potential control behaviours and their effects on the desired control states and the current and desired control domain and worksystem. Colbert's menus, Dowell's list of representations, Norman's plan, and the Hayes-Roths' blackboard model all support the identification of these representations. Table 3.5 shows the potential behaviours on these representations.

The overall ordering is one of monitor→plan→monitor→etc. The behaviours are of the worksystem, humans *and* computers. Adaptive planning (Dowell, 1998) is supported by the regeneration of plans.

Table 3.5: General planning behaviours

Process	Representation
	Monitoring
(re-)generate	Current state of control domain (CDc)
(re-)generate	Current state of control worksystem (CWc)
(re-)generate	Potential control structures and their effect on the desired states (CPSEc)
	Planning
(re-)generate	Desired control domain (CDd)
evaluate	Desired worksystem domain structures (CWd)

Expert and non-expert planning and control behaviours are conceptualised. Non-expert planning and control would require more of the behaviours than expert planning and control. Users of domestic energy management, here, could be experts or non-experts at planning and control.

Planning is similar to design, and the representations are in the terms of design. In current HCI terms, the representations might be understood as "the worksystem's view of the domain and worksystem," rather than that of the designers'.

3.2.4 OPERATIONALISATION

The planning domain and worksystem are operationalised separately from, but related to, the control domain and worksystem. The representation structures are operationalised in the planning domain and worksystem. The process behaviours are operationalised by reference to composite planning and control structures.

Concepts such as "learning," "pre-planning," "reflectiveness of planning," and "meta-planning" are not expected to be operationalised in the development cycle designs.

3.3 CYCLE 1 BEST-PRACTICE DEVELOPMENT

Section 3.1 proposes the development of an artefact specification to satisfy the Cycle 1 user requirements as part of the strategy for acquiring HCI-EDPs. The Cycle 1 user requirements are selected in § 3.2. This section describes the best-practice development of an artefact specification to satisfy the Cycle 1 user requirements.

Best-practice development (specify, implement, test, and evaluate practice, applying best-practice design knowledge available) includes the application of MUSE (Lim and Long, 1994) to the user requirements of the case study.

Informal evaluations of the artefact specification are conducted against the user requirements. The evaluation is positive and the artefact specification appears to satisfy the user requirements. The evaluation means that this Cycle 1 best-practice development supports the operationalisation of the Cycle 1 specific design problem and its solution. The latter is presented in the following section.

3.3.1 USER REQUIREMENTS

The user requirements for Cycle 1 are restated as follows:

> "The domestic routine of *X* occasionally requires them to remain at home to work in the mornings, rather than to leave earlier with his partner, *Z*, to work at the office. However, if *X* leaves after 8 a.m. or stays at home to work, then the house is too cold until they turn the gas-powered central heating back on. If they expect to be at home for a short time after 8 a.m., they often use the one-hour boost facility on the heating controller to turn the heating back on. This boost can result in them being too cold, if they are at home for longer than expected. *X*'s ability to work is adversely affected by being cold and having to control the heating. *X* finds it difficult to plan much in advance, whether they are staying home to work or, if they stay, how long they will stay. The current gas bill is acceptable and an increase could be tolerated, although a decrease would be desirable."

3.3.2 ARTEFACT SPECIFICATION

The artefact specification can best be characterised by comparison with the existing heating controller. The latter is limited to a preset schedule for each day. It is programmed to have two heating "on-off" periods:

- "On" early morning at 6:40 a.m. and "off" at 7:20 a.m.

- "On" early evening at 6:30 p.m. and "off" at 10:00 p.m.

The existing controller does not have appropriate features to meet the user requirements. The best-practice development replaces the existing controller by one that has the facility to:

- switch on in the morning at 6:40 a.m. and switch off at 10:00 p.m. during the week;

- if the heating is turned off during a weekday, using an advance button, then the heating will turn on again in the early evening at 6:30 p.m.;

- remain as the previous controller for the weekends; and

- have an additional remote heating-controller, with an advance button and a bright status light, by the front door.

The occupants of the home need be instructed to use the heating controls as before, except that X should press the advance button on either controller, if the status light is "on" just before leaving to go to work during the week. X is to be considered the user of the designed artefact.

3.3.3 BEST-PRACTICE DEVELOPMENT

Following the strategy, the three phases of the MUSE method and HCI guidelines are applied to the user requirements.

3.3.3.1 Information and Elicitation Phase

The current existing system is analysed in detail. Other existing systems are listed but not analysed. A satisfactory artefact specification results from the first MUSE iteration. Two Task Descriptions are produced. First, a task analysis is conducted, based on an interview in which X introspects about their days (Task Description). Second, X is asked to keep a diary for several mornings, during which they stayed at home and left for work.

These Task Descriptions are generalised (Generalised Task Model of the existing system) to gain an understanding of "generic" mornings (which the design needs to support). The tables for the products for the extant system detail valuable observations, design implications, and speculations that arise. For example, it is observed that X appears to plan using an electronic diary and to-do list. The possibility of interfacing these with the heating control is considered, but dismissed. There is poor correspondence between the departure plan and the electronic diary and to-do list.

The final step develops a task-level conceptual design of the target system (General Task Model of the target system) based on the user requirements and the design implications and speculations, derived from the extant system. The task-level conceptual design documents the design decision to control the heating on departure.

The initial task-level conceptual design suggests a potential for re-use of more detailed extant system features. It was decided to perform a more detailed analysis of the extant system to support that potential. Accordingly, a range of MUSE products are developed that analyse the extant system from its conceptual to its detailed design. For example, the Domain of Design Discourse of the extant system and its System Task Model. Analysis during the Information Elicitation and Analysis phase is the basis of the design in the other phases.

3.3.3.2 Design Synthesis Phase

A text summary of the interaction concerns is constructed (Statement of User Needs), based on the user requirements and the analysis of the extant system. The statement contains:

1. explicit design criteria, such as the need for the artefact cost to be acceptable for the benefits;

2. implicit design criteria, such as the retention of the existing functionality of the controller to support non-weekday-morning tasks;

3. explicit system performance criteria, such as X must not be cold;

4. implicit performance criteria, such as X must be permitted to leave home, when they desire (constraining should not be considered suitable for the artefact specification); and

5. relevant design knowledge, such as an extension of a guideline by Shneiderman (1983) that "human action should be eliminated where no [human] judgement is required" to include "and minimise human action where human judgement is required." This extended guideline confirms the essential task-level decision expressed earlier.

The conceptual design of the conjoint user and computer tasks is advanced (Composite Task Model), maintaining consistency with the accepted foundation of the task-level design developed in the previous phase. Important design decisions are now rationalised—a controller in the same location as the existing one and another controller near the front door.

The design is considered at a lower level of detail by the decomposition of the on-line tasks (System Task Model). At this stage, the guidelines of "transfer of learning," "feedback," and "consistency" (Smith and Mosier, 1986) are applied. For example, transfer of learning is supported by porting effective extant tasks to the target system.

Allocation of function between the user and the artefact is considered. It is difficult, if not impossible, to allocate the user's leaving plan to the controller. So, the controller simply responds to the user's control commands. This allocation corresponds with the HCI guideline that humans are generally better than computers at "drawing on experience and adapting decisions to situations" (Shneiderman, 1983).

The additional remote heating-controller is justified as reminding X to control the heating on leaving.

3.3.3.3 Design Specification Phase

The interaction-level design is advanced (Interaction Task Model and Interface Model). The remote heating-controller is designed with an advance push button to ensure "consistency" between the two controllers. Substantial porting of the existing design is possible, particularly with the layout of the two heating-controllers (Pictorial Screen Layouts).

3.3.4 EVALUATION

Three informal analytic assessments of whether the artefact satisfies the user requirements are conducted, in addition to the assessment of consistency resulting from the application of MUSE. First, an analytic argument is constructed to show that the introduction of the artefact into the home of X and Z should "satisfy" the problem. A form of this analytic argument, commensurate with the user requirements, follows.

> "The proposed artefact should support the domestic routine of X, which occasionally requires them to remain at home to work in the mornings, rather than leave earlier with their partner, Z to work at the office. If X leaves after 8 a.m. or stays at home to work, then the house should remain warm without intervention. The design ensures that the gas-powered central heating remains on rather than turning itself off, which causes X to be uncomfortable, because the house cools. Since the house is no longer too cold, X is not required to turn the heating back on. Therefore, even if X expects to be at home for a short time after 8 a.m. they should not need to use the one-hour boost facility for warmth. X's ability to work should no longer be adversely affected by being cold and having to control the heating. The house is now warm and the heating does not need controlling, until they finish working. X finds it difficult to plan in advance, whether they are staying at work and, if staying, for how long. The artefact should support this planning difficulty, as the heating should only need controlling to match the time of planning. The gas bill may increase by a small amount, which X and Z consider acceptable. X should not be overly taxed by turning the heating off, when leaving, or learning to turn the heating off. The cost of the artefact should be low (approximately £40 for a fully functioning prototype version)."

The second informal analytic assessment involves a panel of nine practitioners, five human factors engineers and four software engineers, appraising the artefact specification resulting from the application of MUSE. They are all familiar with the method and the user requirements. Although some initial objections are raised, after discussion none of these are considered relevant in terms of the artefact satisfying the user requirements. For example, some of the objections either asserted the artefact fulfilled more than the user requirements (but not less) or that the artefact might have embodied alternative design features.

The third, and last, informal analytic assessment is an expert walkthrough of the artefact specification performed by a human factors engineer. Their report contains the following concluding statement.

> "The likely behaviour of the occupants of the house with respect to the system is estimated with respect to a number of scenarios concerning different types of

morning events. It is considered that in the scenario, where there was previously a problem (that is, when X remains at home after 8 a.m.), the system would solve the problem by maintaining X's comfort, and that X would remember to switch the system off, as long as the front door controller is located in a suitably prominent position. When X leaves the house early, their expectations of the system, based on the existing one, may initially cause them to forget to switch the heating off. They are currently not required to take any action, when leaving early in the morning. However, X would soon learn to adapt their morning routine to include the new task of switching the heating off. Similarly, if X left the house earlier than Z, they might forget to switch the heating off, as the normal morning routine does not require any action on Z's part. However, if the system status is designed to be conspicuous, and the controller is prominently located, these problems would be less likely to occur, than if the controller were located in a less visible position. At present, there is no evidence in the user requirements or in the analysis of the existing systems that X will ever leave earlier than Z. Further consultation with X confirms this and so the problem of Z having to remember to operate the system would occur very (and acceptably) infrequently."

In addition, an empirical assessment is performed by constructing a faithful prototype (which does not alter the state of the heating) of the remote heating controller and re-programming the existing controller. This prototype is placed by the front door and the occupants given instruction on its use. This assessment confirms the analytic argument. No empirical assessment of the gas bill increase is completed.

Taken together, the analytic and empirical assessments demonstrate informally that the artefact specification fulfils the user requirements.

3.4   CYCLE 1 OPERATIONALISATION

Section 3.2 proposes the operationalisation of Cycle 1 specific design problem and its solution as part of the strategy for acquiring HCI-EDPs. The Cycle 1 best-practice development is described earlier. This section describes the operationalisation of the Cycle 1 specific design problem and its solution.

The operationalisations are of: the specific design problem and solution conceptions (see § 2.1.4); the conception of human-computer systems (see § 2.3); and the conception of planning and control (See § 3.2). The operationalisation starts with an explicit operationalisation to support the formal and metricated operationalisation. The latter applies the frameworks of § 3.1.

Following the HCI-EDP acquisition strategy, only the cognitive abstract behaviours are operationalised. The conative and affective abstract behaviours and structures are not operationalised.

3.4.1 CURRENT SOLUTION OPERATIONALISATION

The current solution operationalisation is described before the specific design problem and its solution operationalisation.

3.4.1.1 Specific Actual Performance

The planning specific **actual performance** is operationalised as the union of the planning specific actual quality and the planning specific actual costs. The planning **worksystem boundary criteria** are operationalised by the requirement that the constituents of the planning worksystem have the common goals of the current level of achievement and satisfaction of the planning of the comfort of X and the planning of the leaving of X. The planning **domain boundary criteria** are operationalised by the requirement that the constituents of the planning domain of application express the current level of achievement and satisfaction of these common goals.

The control specific **actual performance** is operationalised as the union of the specific actual quality and the specific actual costs. The control **worksystem boundary criteria** are operationalised by the requirement that the constituents of the control worksystem have the common goals of the current (level of) achievement and satisfaction of the control of the comfort of X in the home of X using the heating system and the leaving of X. The control **domain boundary criteria** are operationalised by the requirement that the constituents of the control domain of application express the current (level of) achievement and satisfaction of these common goals.

3.4.1.2 Specific Actual Quality

The planning specific actual domain of application has a main **abstract object** of X's plans, with two **abstract attributes** of leaving plan quality and comfort plan quality. Both of these abstract plan quality attributes have attributes of: time scope; object scope; behaviour scope; view type; view content options; view format options; and content control structures. Each of these plan quality attributes is **related** to the plan quality, and each plan quality is related to the overall plan quality of X's plans. For example, when the planning worksystem finalises a leaving plan, the state of the time scope for the leaving plan changes to indicate when the leaving plan is to occur.

The control specific actual domain of application has two main **physical objects**: X and the study, where X works. X has a **physical attribute** of temperature and an **abstract attribute** of comfort. The attribute of comfort is **related** to the attribute of temperature having a range of acceptable temperatures (between 36.5°C and 37.5°C), when X is in the house. The second main **physical object** is the study, which has a **physical object** of its radiator and a **physical attribute** of the radiator's temperature. The temperature of the study is **related** to the temperature of X, an approximately linear relationship and the temperature of the radiators, related through convection, u-value of the room, etc. The temperature of the radiator is controlled by the worksystem.

The current **states** of the temperatures of the radiators result in the **state** of the comfort attribute of X being "not comfortable," indicated by a "false" Boolean value, at some times. This state is a **task achieved goal** and defines the **product achieved goal** of the **actual quality** by interpretation of the relationships between this attribute and the other attributes in the current actual domain of application.

3.4.1.3 Specific Actual Costs

There are two main sub-systems in the planning worksystem—the planner (X) and the heating controller (a simple two-period time controller). The planner has the **physical behaviour** of feeling the temperature of X. The **abstract behaviours** are mainly contained in the composite behaviours of: standard monitor (type A); standard sub-plan (type A); standard monitor (type B); standard sub-plan (type B); and standard sub-plan (type 0).

The **abstract structures** of the planning worksystem include: the current and desired comfort of X; the current and desired temperature of X; the current and desired location of X, and the time when the heating controller turns off the heating.

There are two main sub-systems in the control worksystem: the user (X) and the heating system (a combination boiler system and the heating controller). The heating system has the following interacting **physical behaviours**: receive press of a one-hour boost button: turn on the LED and turn off the LED. The user has the following interacting **physical behaviour**: perform press of one-hour boost button, and see the LED. The non-interacting **physical behaviours** include, as examples for the heating system, turn the heating on and off, and for the user, walk to and from the location of the heating controller. A further non-interacting **physical behaviour** of the user, and an example of a behaviour that corresponds with the transformation of the attributes of objects in the domain of application, is the closing of the front door, which changes X's "in the house" attribute state to false.

The physical structures are derived from the physical behaviours, for example the heating controller has a physical structure of a one-hour boost button and the user has a physical structure of a body, including a hand that can press and an eye that can see.

The **abstract behaviours** of the heating system include turning off the heating at 7:20 a.m. turning off the heating at the end of the boost period, and the computer operation of addition for the boost timer. The abstract behaviours of the user include forming and popping goals to boost the heating, move to the controller and leave. The **abstract structures** of the heating system are the current boost time and the potential ordering of the heating system behaviours. The abstract structures of the user are the current state of the heating LED and the potential ordering of the user abstract behaviours.

The unitary **behavioural and structural costs** as operationalised over the whole period appear in Table 3.6, for planning, and Table 3.7, for control. The **actual costs** are operationalised by the union of these actual resource costs.

Table 3.6: Planning behavioural and structural costs for Cycle 1 current operationalisation

Main Sub-system	Cost Type	Cost
Planner	Abstract Structural Costs	81
	Physical Structural Costs	1
	Abstract Behavioural Costs	66
	Physical Behavioural Costs	1
Heating System	Abstract Structural Costs	1

Table 3.7: Control behavioural and structural costs for Cycle 1 current operationalisation

Main Sub-system	Cost Type	Cost
User	Abstract Structural Costs	35
	Physical Structural Costs	7
	Abstract Behavioural Costs	41
	Physical Behavioural Costs	11
Heating System	Abstract Structural Costs	16
	Physical Structural Costs	18.9
	Abstract Behavioural Costs	18
	Physical Behavioural Costs	18

3.4.2 SPECIFIC DESIGN PROBLEM OPERATIONALISATION

The desired operationalisation aims for a minimal expression, which is achieved by using quality and costs statements with respect to the current operationalisation.

3.4.2.1 Specific Desired Quality

The main **task goal** is to maintain the state of X's comfort attribute as "comfortable" instead of a task achieved goal of "not comfortable." The comfort plan quality should be acceptable. The leaving plan quality should also be acceptable, including permitting X to leave, when they wish.

3.4.2.2 Specific Desired Costs

The **physical structural costs** of the heating system should be within a range that allows for the preferred decrease or an acceptable increase in gas and electricity usage. It is assumed that the heating system can be modified and, therefore, the operationalisation of the **physical and abstract structural costs** of the heating system should be within a range that allows for a different installation and maintenance price. Further, it is expected that a small increase in **physical and abstract behavioural costs** of the heating system would be acceptable and this increase would be reflected in the operationalisation within a range of acceptable costs. It is assumed that the user costs either remain the same, or decrease if possible.

3.4.3 SPECIFIC DESIGN SOLUTION OPERATIONALISATION

The current solution operationalisation is described after the specific design problem and its solution operationalisation.

3.4.3.1 Specific Actual Performance

The planning specific **actual performance** is operationalised as the union of the planning specific actual quality and planning specific actual costs. The planning worksystem criteria are operationalised by the requirement that the constituents of the planning worksystem have the common goals of the actual (level of) achievement and satisfaction of the planning of the comfort of X and the leaving of X. The planning **domain boundary criteria** are operationalised by the requirement that the constituents of the planning domain of application express the actual (level of) achievement and satisfaction of these common goals.

The control specific **actual performance** is operationalised as the union of the control specific actual quality and the control specific actual costs. The control **worksystem criteria** are operationalised by the requirement that the constituents of the planning worksystem have the common goals of the actual (level of) achievement and satisfaction of the control of the comfort of X in the home of X using the heating system and the leaving of X. The control **domain boundary criteria** are operationalised by the requirement that the constituents of the control domain of application express the actual (level of) achievement and satisfaction of these common goals.

3.4.3.2 Specific Actual Quality

The planning and control domains of application are the same as those in the current operationalisation. The **task achieved goal** is that the state of the comfort attribute of X is "comfortable" (true) for all times, as expected by a solution. This state is achieved through the **state** of the temperature **attribute** of X being held between the range of acceptable temperatures for X's comfort. The **state** of the temperature of the study is held relatively constant by the **state** of the temperatures of the radiator. All of these states describe the **product achieved goal**.

3.4.3.3 Specific Actual Costs

There is one main sub-system in the planning worksystem—the planner (X). The planner has the **physical behaviour** of seeing the heating system LED. There are fewer, relative to the current, occurrences of the composite **abstract behaviours**. The **abstract structures** of the planning worksystem remain the same.

There are two main sub-systems in the control worksystem—the user (X) and the heating system (a combination boiler system and a simple two-period time controller with remote advance controller). The heating system has the following interacting **physical behaviours**: receive press of front-door advance button and turn off the LED. The user has the following interacting **physical behaviours**: perform press of front-door advance button, and see the LED. Examples of **physical structures** are, for the heating system, a front-door advance button and, for the user, a hand that can press.

The **abstract behaviours** for the heating system include turning off the heating on the advance button. The abstract behaviours of the user include forming and popping goals to leave and advance the heating. The **abstract structures** of the heating system are the current advance state and the potential ordering of the heating system behaviours. The abstract structures of the user are the current state of the heating LED and the potential ordering of the user abstract behaviours.

The behavioural and structural costs as operationalised over the whole period are presented in Table 3.8, for planning, and in Table 3.9, for control. The **actual costs** are operationalised by the union of these actual resource costs.

Table 3.8: Planning behavioural and structural costs for Cycle 1 solution operationalisation

Main Sub-system	Cost Type	Cost
Planner	Abstract Structural Costs	81
	Physical Structural Costs	1
	Abstract Behavioural Costs	66
	Physical Behavioural Costs	1
Heating System	Abstract Structural Costs	1

52 3. CYCLE 1 DEVELOPMENT OF INITIAL HCI ENGINEERING DESIGN PRINCIPLES

Table 3.9: Control behavioural and structural costs for Cycle 1 solution operationalisation

	Cost Type	Cost
User	Abstract Structural Costs	35
	Physical Structural Costs	7
	Abstract Behavioural Costs	41
	Physical Behavioural Costs	11
Heating System	Abstract Structural Costs	16
	Physical Structural Costs	18.9
	Abstract Behavioural Costs	18
	Physical Behavioural Costs	18

REVIEW

The chapter presents conceptions to support operationalising specific design problems and solutions, including the concept of composite structures. An initial conception of planning and control is proposed to operationalise the subsequent design cycles. The resulting artefact specification is considered to satisfy the Cycle 1 user requirements. Cycle 1 best-practice development provides the basis for the operationalisation of the specific design problem and its solution.

3.5 PRACTICE ASSIGNMENT

3.5.1 GENERAL

Read § 3.1, concerning the operationalisation of specific design problems and solutions.

- Check the operationalisation informally for completeness and coherence, as required by the case study of domestic energy planning and control.

- The aim of the work assignment is for you to become sufficiently familiar with the operationalisations to apply them subsequently and as appropriate to a different domain of application, as in Practice Scenario 3.1.

Hints and Tips

Difficult to get started?

Re-read the assignment task carefully.

- Make written notes and in particular list the sections, while re-reading § 3.1.

- Think about how the sections might be applied to describe the operationalisation of specific design problems and solutions of a novel domain of application.

- Re-attempt the assignment.

Test

List from memory as many of the sections as you can.

Read § 3.2, concerning the conception of planning and control, as it relates to the application domain of domestic energy planning and control.

- Complete as for the previous section beginning Read § 3.1.

Read § 3.3, concerning the Cycle 1 best-practice development, as it relates to the application domain of domestic energy planning and control.

- Complete as for Read § 3.1.

Read § 3.4, concerning the Cycle 1 operationalisation, as it relates to the application domain of domestic energy planning and control.

- Complete as for Read § 3.1.

3.5.2 PRACTICE SCENARIO

Practice Scenario 3.1: Operationalising Cycle 1 in an Additional Domain of Application

Select an additional domain of application.

- Apply the operationalisation of specific design problems and solutions for the domestic energy planning and control (see § 3.1) to the novel domain of application. The description can only be of the most general kind—that is at the level of operationalisation. However, even consideration at this high level can orient the researcher towards application of the operationalisation to novel domains of application. The latter are as might be required subsequently by their own work. The practice scenario is intended to help bridge this gap.

- Operationalise Cycle 1 for the additional domain.

CHAPTER 4

Cycle 2 Development of Initial HCI Engineering Design Principles for Domestic Energy Planning and Control

SUMMARY

This chapter reports the development of initial HCI-EDPs for the application domain of domestic energy planning and control. This chapter comprises: Cycle 2 best-practice development, and Cycle 2 operationalisation.

4.1   CYCLE 2 BEST-PRACTICE DEVELOPMENT

Section 2.2 proposes the development of an artefact specification to solve Cycle 2 user requirements as part of the strategy for acquiring HCI-EDPs. Cycle 2 user requirements are selected in § 2.2.7.2. The present section describes the best-practice development of an artefact specification to satisfy Cycle 2 user requirements. As earlier, best-practice development is taken to include the application of MUSE (Lim and Long, 1994) to the user requirements.

Informal evaluations are presented of the artefact specification against the user requirements. The evaluation is positive. The artefact specification satisfies the user requirements. The evaluation indicates that Cycle 2 best-practice development supports the operationalisation of Cycle 2 specific design problem and its solution.

4.1.1   USER REQUIREMENTS

The user requirements for Cycle 2 are restated as follows.

> "The kitchen is usually a very comfortable room, probably because it has thick walls. However, it can get too hot when P is cooking, even in the winter. The room has three radiators with individual thermostats. The radiators are heated, using hot water from a gas-powered combination boiler in another room. There is no central thermostat for the boiler, but there is a time-controller and a water temperature controller, neither of which are in the kitchen. The boiler supplies other radiators in the house. There is an extractor fan over the cooker, but it is broken. Double-glazed

4.1.2 ARTEFACT SPECIFICATION

The artefact specification provides additional cooling and support for *P*'s cooking and cooling planning and control. A new fan is specified to provide additional cooling, rather than repair the broken extractor-fan. Even if mended, the extractor-fan would not provide significant cooling.

The planning support for the cooking activities and the heating is a pre-printed A3 surface covered in laminated plastic. *P* writes on it with a water-soluble pen, so changes can be made, including starting a new plan. Two pens are available, one for planning and the other for re-planning during control. The pre-printing provides prompts and space for an explicit representation of the plans and some of their criteria.

A controller is provided for the door, the fan, and the radiators in the kitchen. The controller permits entry and display of the heating plan as it relates to cooking periods. A pre-printed, laminated booklet supports the documentation of previous times of cooking activities to aid the planning for the cooking activities. Instructions are printed on the front of the booklet.

4.1.3 BEST PRACTICE DEVELOPMENT

Following the research strategy, the MUSE method phases and HCI guidelines, as best-practice at the time of the research, are applied to the user requirements.

4.1.3.1 Information and Elicitation and Analysis Phase

The current existing system is analysed in detail. Two Task Descriptions are produced. Three scenarios are elicited by paper-based questioning. All of the scenarios involve meals that resulted in *P* becoming too hot. *P* was observed and questioned, concerning their range of cooking tasks, including meal planning, shopping, and cooking planning.

These Task Descriptions are generalised (Generalised Task Model of the existing system). Useful observations, design implications, and speculations arose during this phase. For example, it was recognised that the kitchen door provided effective cooling, but that it was not often opened. This suggests that the kitchen door should be opened more often for cooling. Also, *P* became "flustered" during cooking, contributing to *P* becoming too hot. This suggests that a reduction in *P* becoming flustered would reduce *P* becoming too hot.

The initial task-level conceptual design of the target system (General Task Model of the target system) documented: the essential design decisions for more and earlier planning of the cooking; early planning of the heating; turning off the heating, even in winter if necessary; support for re-planning during cooking; and support for improving future planning.

4.1 CYCLE 2 BEST-PRACTICE DEVELOPMENT   57

As in Cycle 1, analysis during the Information Elicitation and Analysis Phase is the basis for the design in the other phases.

4.1.3.2 Design Synthesis Phase

A textual summary of the HCI concerns (Statement of User Needs) included: any explicit design criteria, such as the amount of fuel used cannot increase very much, and desirably would decrease. Also, any implicit design criteria, such as the artefact cost should be low, any explicit system performance criteria, such as P must not be too hot. Further, any implicit performance criteria, such as P must be able to cook the meals that she desires, when she wishes. Last, any relevant human factors knowledge, as best-practice, such as feedback and consistency guidelines should be followed.

The conceptual design of the conjoint user and computer tasks is advanced (Composite Task Model). Important design decisions are rationalized: the provision of an additional fan; when the door should be opened, the fan turned on, and the radiators turned off. Also, the explicitness of the cooking and heating plan. The desired increase in meal and heating planning suggests support for both types of planning. A controller is rationalised to off-load the control during cooking.

The on-line tasks are decomposed (System Task Model) to support the ordering of cooking and heating planning. Internal iteration produces a "bubbled up" rationale for two devices—one for planning and another for control. Tasks are identified for their support and the transfer of information from the planning support to the control support.

4.1.3.3 Design Specification Phase

The interaction-level design is advanced (Interaction Task Model and Interface Model). A paper-based planning and memory aid, covered in laminated plastic, is rationalised to support being cleaned and amended during use. The latter is facilitated by the use of water-soluble markers, being carried around, both in the kitchen and out. Also, being stood-up in the kitchen and being lightweight and available.

Computer-supported planning is rejected. Putting a computer in the kitchen would be inconvenient and P has a dislike of electronic gadgets. Lines are rationalised for showing the planned length of cooking activities, with their thickness indicating the effort.

An electronic controller is rationalised to control the radiators in the kitchen and the fan. Also, to remind P when to open the door. Consistency between the planning aid and the controller improves transfer between the devices.

The "screen" layouts of the planning aid and controller are designed (Display Design).

58 4. CYCLE 2 DEVELOPMENT OF INITIAL HCI ENGINEERING DESIGN PRINCIPLES

4.1.3.4 Evaluation

Three informal analytic assessments of whether the artefact satisfies the user requirements are conducted. This is in addition to the assessment of consistency, resulting from the application of MUSE.

First, an argument is constructed, showing that the artefact should remove P's problem. A form of the argument, commensurate with the user requirements, follows.

> "The artefact should support P in improved planning of meals, the activities involved in generating meals, and the required heating. Improved planning of the meals and their activities should prevent P from becoming flustered, during meal preparation. Improved planning of the heating should enable P to control the heating so that P will be kept cool at the correct times during cooking."

The second analytic assessment involved a panel of seven practitioners, five human factors engineers and two software engineers, appraising the artefact specification produced using MUSE. They were all familiar with the method and the user requirements. No objections were maintained, such that the artefact failed to satisfy the user requirements.

The third analytic assessment was an expert walkthrough of the artefact specification performed by a human factors engineer. Their report contained the following concluding statement.

> "Based on my examination of the meal planning aid and my discussions with P, it is my opinion that use of the planning aid is likely to result in improved meal planning and less heat in the kitchen at busy points during meal preparation. P should therefore avoid becoming flustered and too hot. Due to the effort involved in planning, I anticipate that the sheet will probably only be required for more complicated meals. It is reasonable to expect that use of the sheet on these occasions will result in a "transfer of training" to simple meals. Also, that P's awareness of the need for ventilation and cooling of the kitchen will be improved as a result of using it. Initially, I was concerned that use of the planner would be abandoned during busy periods in the kitchen, exactly when it is required most. However, P appears to be of a very methodical nature, always planning meals well in advance and preparing and using a detailed shopping list. Given P's existing use of lists, and the apparent satisfaction, derived from making and executing plans, I would expect P to find using the meal planner during busy periods both natural and easy."

In addition, an empirical assessment has been performed by constructing an interactively faithful prototype of the planning aid and the controller. The prototype was employed in cooking a complicated meal that would normally be expected to cause P to become too hot. P was less hot, and was not flustered. Minor changes were proposed, for example using numbers instead of lines to represent timing and effort. In a second empirical assessment, P was not hot. This assessment confirms the analytic argument.

Taken together and as best practice, the analytic and empirical assessments demonstrate, that the artefact specification satisfies the user requirements.

4.2 CYCLE 2 OPERATIONALISATION

Section 4.2 proposes the operationalisation of the Cycle 2 specific design problem and its solution, as part of the strategy for acquiring HCI-EDPs. Cycle 2 best-practice development is described in the previous section. Operationalisation of Cycle 2 specific design problem and its solution are described next.

The operationalisations are of: the specific design problem and solution conceptions (see § 2.1.4); the conception of human-computer systems (see § 2.3); and the conception of planning and control (see § 3.2). The operationalisation starts with an explicit operationalisation (similar to the brief description here) to support the formal and metricated operationalisation.

A video of the cooking, with concurrent verbal protocol, is analysed to support the operationalisation.

4.2.1 GENERALITY CONCERN

Due to uncontrollable circumstances (a broken leg), P was not able to provide access for the operationalisation. As a result, the researcher (X) recreated the conditions to support the operationalisation. He cooked one of P's recipes with and without the planning sheet in a similar environment to P. He became hot and flustered without the planning sheet, but was not hot or flustered with the planning sheet. The cooking was conducted in the kitchen in his home. The kitchen has a window and a back door. A fan was fitted.

4.2.2 CURRENT SOLUTION OPERATIONALISATION

The current solution operationalisation is described before the specific design problem and its solution operationalisation.

4.2.2.1 Specific Actual Performance

The planning specific **actual performance** is operationalised as the union of the planning specific actual quality and the planning specific actual costs. The planning **worksystem boundary criteria** are operationalised by the requirement that the constituents of the planning worksystem have the common goals of the current (level of) achievement and satisfaction of the planning of the cooking by X and the heating by X. The planning **domain boundary criteria** are operationalised by the requirement that the constituents of the planning domain of application express the current (level of) achievement and satisfaction of these common goals.

The control specific **actual performance** is operationalised as the union of the specific actual quality and the specific actual costs. The control worksystem boundary criteria are operationalised by the requirement that the constituents of the control worksystem have the common goals of the current (level of) achievement and satisfaction of the control of the cooking by X. Also, the heating by X, in the kitchen of X, using the kitchen's cooker, radiators, and door. The **control domain boundary criteria** are operationalised by the requirement that the constituents of the control domain of application express the current (level of) achievement and satisfaction of these common goals.

4.2.2.2 Specific Actual Quality

The planning specific actual domain of application has a main **abstract object** of X's plans, with two **abstract attributes** of cooking plan quality and heating plan quality. These two plan quality attributes both have attributes of: time scope; object scope; behaviour scope; view type; view content options; view format options, and content control structures. Each plan quality attribute is **related** to the plan quality. The latter is related to the overall plan quality of X's plans. For example, when the planning worksystem finalises a cooking plan the state of content structure changes to reflect the next ingredient required for the cooking.

The control specific actual domain of application has two main **physical objects**, X and the kitchen. There is one main **abstract object** of the meal. X has **physical attributes** of temperature and activity, which are **related** to the abstract attributes of comfort and agitation. The kitchen has **physical objects** of the cooker, radiators, and the door. The **physical attributes** of the temperature of the cooker, the temperature of the radiators, and the airflow of the door are related to the abstract attribute of the temperature of the kitchen, which is related to X's temperature. The temperature of the cooker and door are controlled by the worksystem. The meal has an abstract attribute of quality, which is related to its physical attributes of flavour, presentation, and location.

The current **states** of the door's airflow and the temperature of the cooker result in: the **state** of the comfort **attribute** of X being "not comfortable" (false); the **state** of the agitation **attribute** of X being "agitated" (a high percentage), and the **state** of the quality **attribute** of the meal being "poor," with a value of 7.3, at some times. These states are **task achieved goals** and define the **product achieved goal** of the **actual quality** by interpretation of the relationships between this attribute and the other attributes in the actual domain of application.

4.2.2.3 Specific Actual Costs

There is one main sub-system in the planning worksystem, that is the planner (X). The planner has the **physical behaviour** of seeing the current ingredients used in the cooking. The **abstract behaviours** are contained in the composite behaviours of standard monitor (Type 0) and standard sub-plan (Type 0).

The abstract structures of the planning worksystem include: the current and desired ingredients of the meal, and the current and desired temperature of X.

The main sub-systems in the control worksystem are the user (X) and the cooker. The cooker has the interacting **physical behaviour** of change the level of the gas ring or the oven (a composite behaviour). Correspondingly, the user has the interacting **physical behaviours** of change the level of the gas ring and the oven.

The **physical structures** can be derived from the physical behaviours, for example the cooker has a physical structure of a gas ring, and the user has a physical structure of a hand (that can change the level of the gas of a ring).

The *abstract behaviours* of the cooker include increasing the ring/oven heat with a clockwise turn of a knob. The abstract behaviours of the user include forming and popping goals to make lasagna, cook onions, collect pasta from the cupboard, and assemble lasagna.

The **behavioural and structural costs** as operationalised over the whole period are in Table 4.1 for planning, and Table 4.2, for control. The **actual costs** are operationalised by the union of these actual resource costs.

Table 4.1: Planning behavioural and structural costs for Cycle 2 current operationalisation		
Main Sub-system	**Cost Type**	**Cost**
Planner	Abstract Structural Costs	91
	Physical Structural Costs	2
	Abstract Behavioural Costs	214
	Physical Behavioural Costs	5

Table 4.2: Control behavioural and structural costs for Cycle 2 current operationalisation

Main Sub-system	Cost Type	Cost
User	Abstract Structural Costs	256
	Physical Structural Costs	65
	Abstract Behavioural Costs	382
	Physical Behavioural Costs	70
Heating System	Abstract Structural Costs	90
	Physical Structural Costs	33.48
	Abstract Behavioural Costs	114
	Physical Behavioural Costs	38

4.2.3 SPECIFIC DESIGN PROBLEM OPERATIONALISATION

The desired operationalisation aims for a minimal expression, which is achieved by using quality and costs statements with respect to the current operationalisation.

4.2.3.1 Specific Desired Quality

The main **task goal** is to maintain the state of: X's comfort attribute as "comfortable"; X's agitation attribute as "not agitated," and the meal's quality attribute as "good."

4.2.3.2 Specific Desired Costs

The **physical structural costs** of the heating system should be within a range that allows for a desirable decrease or acceptable increase in gas and electricity usage. It is assumed that the heating system can be modified and, therefore, the operationalisation of the **physical and abstract structural costs** of the heating system should be within a range that allows for a different installation and maintenance price. A small increase in **physical and abstract behavioural costs** of the heating system would be tolerated. This increase would be reflected in the operationalisation within a range of acceptable costs. It is assumed that the **user costs** either remain the same or decrease, if possible.

4.2.4 SPECIFIC DESIGN SOLUTION OPERATIONALISATION

The current solution operationalisation is described after the specific design problem and its solution operationalisation.

4.2.4.1 Specific Actual Performance

The planning specific **actual performance** is operationalised as the union of the planning specific actual quality and the planning specific actual costs. The planning **worksystem criteria** are operationalised by the requirement that the constituents of the planning worksystem have the common goals of the actual (level of) achievement and satisfaction of the planning of the cooking by X and the heating by X. The planning **domain boundary criteria** are operationalised by the requirement that the constituents of the planning domain of application express the actual (level of) achievement and satisfaction of these common goals.

The control specific **actual performance** is operationalised as the union of the control specific actual quality and the control specific actual costs. The control **worksystem criteria** are operationalised by the requirement that the constituents of the planning worksystem have the common goals of the actual (level of) achievement and satisfaction of the control of the cooking by X and the heating by X in the kitchen of X using the kitchen's cooker, radiators, door, and fan. The control **domain boundary criteria** are operationalised by the requirement that the constituents of the control domain of application express the actual (level of) achievement and satisfaction of these common goals.

4.2.4.2 Specific Actual Quality

The planning and control domains of application are the same as those in the current operationalisation. During development, an iteration was required to ensure that the current (and, therefore, problem) operationalisation domain of application is the same as that for the solution application.

The specific actual quality has a **task achieved goal** such that the state of: X's comfort attribute is "comfortable"; X's agitation attribute "not agitated" and the meal's quality attribute "good." The states of these attributes are achieved by: the state of X's temperature attribute being held between the range of acceptable temperatures for X's comfort; the rate of change in X's activity being low, and the state of the meal's flavour, presentation, and location being tasty, well presented, and on the table, respectively. All of these states describe the **product achieved goal**.

4.2.4.3 Specific Actual Costs

There are two main sub-systems in the planning worksystem. They are the planner (X) and the planning-aid. The planner has the **physical behaviour** of seeing the current ingredients, and, for the solution, seeing and writing on the planning-aid. The planning-aid has physical behaviours of displaying and accepting writing. The **abstract behaviours** of the planner are contained in the composite behaviours of: sheet sub-plan (Type 0); standard monitor (Type 0); and standard sub-plan (Type 0).

The abstract structures of the planning worksystem include: the current and desired ingredients of the meal, and the current and desired temperature of X.

There are four main sub-systems in the control worksystem: the user (X); the cooker; the door; and the fan. (The radiators are turned off in summer and the heating controller is not used in this analysis.) The fan has the interacting **physical behaviour** of accept button press to turn on. Correspondingly, the user has the interacting **physical behaviours** of press button to turn on the fan.

The **behavioural and structural costs** as operationalised over the whole period are in Table 4.3, for planning, and Table 4.4, for control. The **actual costs** are operationalised by the union of these actual resource costs.

Table 4.3: Planning behavioural and structural costs for Cycle 2 solution operationalisation

Main Sub-system	Cost Type	Cost
Planner	Abstract Structural Costs	160
	Physical Structural Costs	5
	Abstract Behavioural Costs	1,232
	Physical Behavioural Costs	51
Heating System	Abstract Structural Costs	3
	Physical Structural Costs	4
	Abstract Behavioural Costs	35
	Physical Behavioural Costs	35

Table 4.4: Control behavioural and structural costs for Cycle 2 solution operationalisation

Main Sub-system	Cost Type	Cost
User	Abstract Structural Costs	300
	Physical Structural Costs	75
	Abstract Behavioural Costs	488
	Physical Behavioural Costs	125
Heating System	Abstract Structural Costs	90
	Physical Structural Costs	33.26
	Abstract Behavioural Costs	102
	Physical Behavioural Costs	34

REVIEW

This chapter describes how Cycle 2 best-practice development supports the operationalisation of the specific design problem and its solution. The artefact specification is considered to satisfy the Cycle 2 user requirements.

4.3 PRACTICE ASSIGNMENT

4.3.1 GENERAL

Read § 4.1, concerning the Cycle 2 best practice development.

- Check the best practice development informally for completeness and coherence, as required by the case study of domestic energy planning and control.

The aim of the practice assignment is for you to become sufficiently familiar with the best practice development to apply it subsequently, and as appropriate, to a different domain of application, as in Practice Scenario 4.1.

Hints and Tips

Difficult to get started?

Re-read the assignment task carefully.

- Make written notes and in particular list the sections, while re-reading § 4.1.

- Think about how the sections might be applied to describe Cycle 2 best practice and operationalisation in a novel domain of application.

- Re-attempt the assignment.

Test

List from memory as many of the sections as you can.

4.3.2 PRACTICE SCENARIO

Practice Scenario 4.1: Operationalising Cycle 2 in an Additional Domain of Application

Select an additional domain of application.

- Apply the operationalisation of Cycle 2 for the domestic energy planning and control (see § 4.1) to the novel domain of application. The description can only be of the most general kind—that is at the level of operationalisation. However, even application at this high level can orient the researcher towards the operationalisation of novel domains. The latter are as might be required subsequently by their own work. The research design scenario is intended to help bridge this gap.

CHAPTER 5

Initial HCI Engineering Design Principles for Domestic Energy Planning and Control

SUMMARY

This chapter proposes an instance-first strategy for the acquisition of HCI-EDPs. When applied, the strategy produces early, initial such principles for the application domain of domestic energy planning and control. The principles derive from a range of origins.

5.1 DETAILED STRATEGY

Almost any part of a specific design problem and its solution is a potentially general relationship. However, it is possible to investigate those more likely to be general than others. The following six forms of the latter are detailed later, with examples from the cycle operationalisations: initial HCI-EDPs, identified during operationalisation(s); initial assumption assessment from operationalisation(s); inspirational initial such principles from operationalisation(s); initial such principles from general guidelines; initial such principles from MUSE guidelines; and initial such principles from MUSE tasks (Lim and Long, 1994).

The operationalisation of the current solution is included in the research strategy to support operationalisation of the specific design problem. However, the (minimalist) specific design problem operationalisations require the current solution operationalisations to make sense (where actual performance is less than desired performance). The operationalisations of the current solution, then, are considered part of the operationalisations of the specific design problem.

Some generality is acquired by identifying commonalities between the operationalisations, termed "inter-initial engineering design principles" (see later). Following the operationalisations, initial HCI-EDPs might be general *within* a particular operationalisation, termed "intra-initial principles." The latter probably have a higher likelihood of being HCI-EDPs than those with no such generality, but with a lower likelihood than inter-initial such principles. Intra-initial HCI-EDPs can also be understood by considering the specific design problem and its solution as containing specific design sub-problems and their solutions, which have intra-initial HCI-EDPs.

68   5. INITIAL HCI ENGINEERING DESIGN PRINCIPLES FOR DOMESTIC ENERGY

However, with this understanding, their sub-problem basis probably rests on it being part of an overall problem. Both types of principle are exemplified.

5.1.1   GENERALITY OF THE INITIAL HCI ENGINEERING DESIGN PRINCIPLES

The initial HCI-EDPs have generality through:

- commonalities, from parameterisation and "null components," where concepts are not operationalised in the initial HCI-EDPs. Null components indicate generality over the concepts not operationalised. For example, a null problem component suggests generality over all problems of the cycle types;

- composite structures; and

- cycle types, which arise from the earlier assertion that "a general design problem and its general design solution are general over types of user, types of computer and types of domain of application."

Types for each category follow.

1. Cycles 1 and 2 User Types

   The user in both the cycles is X. The types of user for X include: researcher, male, aged 32, postgraduate, etc. This list of types could be developed. Neale and Liebert (1986) suggest further context concerns or "external validity"; "population validity"; "geographic areas validity"; "temporal validity"; and "[designer] validity."

2. Cycle 1 Heating Controller Types.

   Types of heating controller include: a simple controller; a two-period controller; a heating controller; a domestic heating controller; a domestic energy management system; an energy management system, etc.

3. Cycle 1 System Heating Types.

   Types of heating system include: a combination boiler heating system; a gas-powered heating system; an energy delivery system; etc.

4. Cycle 2 Cooker Types.

   Types of cooker include: an upright cooker; a gas cooker; a domestic cooker, etc.

5. Cycle 1 Domain Types.

Types of domain include: comfort planning and control; leaving planning and control; domestic energy management; energy management; late comfort planning; late leaving planning; etc.

6. Cycle 2 Domain Types.

Types of domain include: comfort planning and control; cooking planning and control; domestic energy management; energy management; late comfort planning; minimal cooking planning; etc.

5.1.2 GENERALISATION OVER TYPES

Inter-initial HCI-EDPs require generalisation over the above types. Generalisation occurs in two ways. First, types that are common to both cycles are carried forward to the inter-initial engineering principle's types. Second, the power set of types that are not common to both cycles, are carried forward to the inter-initial engineering design principle's types.

5.2 INITIAL HCI ENGINEERING DESIGN PRINCIPLES IDENTIFIED DURING OPERATIONALISATION(S)

The closest targeting to the original strategy is to "identify" initial HCI-EDPs both within operationalisations (intra-initial such principles) and across operationalisations (inter-initial such principles). These initial principles may be in the formulae or in the values. "Identify" in this case is an iterative search.

In the values, the initial HCI-EDPs may be across, down, or both (in the senses of the spreadsheet), as concerns the operationalisation. Across relates to the initial HCI-EDPs between the changes of behaviours, structures, and states of the domain. Down relates to the changes for a particular behaviour, structure, or state of the domain.

In the formulae examples, which follow, any X preceding a colon (that is, "X:") always and only refers to the user of the system (as in § 5.1.1). All remaining Xs are to be interpreted, following the conventions of the formulae. User X is to maintain consistency with such references elsewhere in the book.

5.2.1 EXAMPLES

Examples 1–3 are examples of initial HCI-EDPs identification in Cycle 1 and Cycle 2 operationalisations. They show that almost any part of a specific problem and solution can be considered a relationship.

Example 4 is an example of an intra-initial HCI-EDP, identified in Cycle 1 operationalisation.

Example 5 is an example of an inter-initial HCI-EDP, identified in Cycle 1 and Cycle 2 operatinalisations.

- Example 1 in Operationalisation 1

In the current system, StMonA is always followed directly by StSubPlanA. The formulae show this outcome as (for example):

| F11 | X:StMonA:FP, FeelTemp, Temp, Comfort | |
| G11 | X:StSubPlanA:RP, In house, Comfort | =F10 |

This situation is common to many of the formulae. This initial HCI-EDP can be shown in a notation:

$$X : \text{StMonA} : \text{X,Y,Z} \xrightarrow[1,e=1]{} X : \text{StSubPlanA} : \text{P,Q,R}$$

where the arrow shows the "followed by" relationship, the first number under the arrow (1) shows the likelihood (probability) of the follow relationship, and the second number under the arrow (e,=1) shows the number of event ticks in the follow relationship. The double arrow down shows the direction of design.

Similarly,

$$X : \text{StMonB} \xrightarrow[1,e=1]{} X : \text{StSubPlanB}$$

- Example 2 in Operationalisation 1

In the actual system:

$$X : \text{StMonA} : \text{X, Y, Z} \xrightarrow[1,e=1]{} X : \text{StSubPlanA} : \text{P, Q}$$

Combining this initial HCI-EDP with the one above results in, for example, the following initial HCI-EDP:

$$X : \text{StMonA} : \text{X, Y, Z} \xrightarrow[1,e=1]{} X : \text{StSubPlanA} : \text{P, Q, R}$$
$$\Downarrow$$
$$X : \text{StMonA} : \text{A, B, C} \xrightarrow[1,e=1]{} X : \text{StSubPlan} : \text{M, N}$$

where, the double down arrow shows the direction of design, starting at the current system and ending at the actual system. It is possible to relate the capital letters for arguments, where they represent increased generality:

$$X : \text{StMonA} : \text{X,Y,Z} \xrightarrow[1,e=1]{} X : \text{StSubPlanA} : \text{P,Q,Z}$$
$$\Downarrow$$
$$X : \text{StMonA} : \text{X,B,Z} \xrightarrow[1,e=1]{} X : \text{StSubPlan} : \text{P,Z}$$

- Example 3 in Operationalisation 1

Including the domain state changes with the above behavioural changes results in:

$$X : \text{StMon A} : X, Y, Z \qquad\qquad X : \text{StSubPlanA} : P, Q, Z$$

$$\downarrow 1,0 \qquad \xrightarrow{1,e=1} \qquad \downarrow 1,0$$

$$X : \text{CDc} : \text{Current } Z = \text{FALSE} \qquad X : \text{CDd} : \text{Desired } Z = \text{TRUE}$$
$$\Downarrow$$

$$X : \text{StMon A} : Z, B, Z \qquad\qquad X : \text{StSubPlan} : P, Z$$

$$\downarrow 1,0 \qquad \xrightarrow{1,e=1} \qquad \downarrow 1,0$$

$$X : \text{CDc} : \text{Current } Z = \text{TRUE} \qquad X : \text{CDd} : \text{Desired } Z = \text{TRUE}$$

where the down arrow shows the domain state change that follows the behaviour occurring. The first number beside the arrow (1) shows the likelihood (probability) of the follow relationship, and the second number under the arrow (2) shows the number of event ticks in the follow relationship.

- Example 4: Intra Operationalistion 2.

The following initial principle holds 22 times within Operationalisation 2. The space before the double arrow shows that there are no general components of the current system in this SR.

$$\Downarrow$$
$$X : \text{ShSubPlan} : X, \text{Ingredient} \xrightarrow{1,e=1} X : \text{ShSubPlan} : X, \text{Ingredient}$$

Similarly, there is another initial HCI-EDP that indicates much the same information, over the whole of Operationalisation 2:

$$\Downarrow$$
$$X : \text{ShSubPlan} : X, \text{Ingredient} \xrightarrow{1,e=1} X : \text{ShSubPlan} : X, \text{Ingredient}$$

- Example 5: Inter Operationalisation 1 and Operationalisation 2

In Operationalisation 1:

$$\Downarrow$$
$$X : \text{StSubPlan} : RP, \text{Comfort}$$
$$\downarrow$$
$$X : \text{CDc} : \text{Desired Comfort} = \text{TRUE}$$

In Operationalisation 2:

$$\Downarrow$$
$$X : \text{ShSubPlan} : \text{FS, Comfort}$$
$$\downarrow$$
$$X : \text{CDc} : \text{Desired Comfort} = \text{TRUE}$$

Generalising the StSubPlan and ShSubPlan composites to a StShSubPlan: X, Y, Z, where Z can be Store or Write, leads to a general inter-initial HCI-EDP (over Operationalisation 1 and Operationalisation 2) of:

$$\Downarrow$$
$$X : \text{StShSubPlan} : \text{FS, Comfort, Z}$$
$$\downarrow$$
$$X : \text{CDc} : \text{Desired Comfort} = \text{TRUE}$$

5.3 INITIAL ASSUMPTION ASSESSMENT FROM OPERATIONALISATION(S)

The initial assumption is that the underlying conceptions can be assessed, given that they have been operationalised successfully. Since these initial assumptions are intended to be general, they provide a basis for generality. For example, the "monitor→plan→monitor" conception of planning and control appears in both operationalisations and so is general over both.

5.3.1 EXAMPLES

Since the "monitor→plan→monitor" conception of planning and control is general over both operationalisations it can be assessed.

"Monitor" is considered to be an StMon or an StMonA behaviour, and "plan" to be an StSubPlanA, an StSubPlanB, an StSubPlan, or an ShSubPlan behaviour. The following example corollaries might result in generalities:

- Example 1: If there is a monitor behaviour then it will be followed by a plan behaviour.

$$\text{Monitor} \xrightarrow{1, e \geq 1} \text{Plan}$$

This condition holds in both operationalisations so it is general over both.

- Example 2: If there is a plan behaviour, then it will always be followed by a monitor behaviour.

$$\text{Plan} \xrightarrow{1, e \geq 1} \text{Monitor}$$

This condition is violated in the Operationalisation 1 current and actual solution and the Operationalisation 2 current and actual solutions. (Violated because the conception is unreasonable—since planning must stop at some point, but also violated because the conception does not explicitly allow for planning without monitoring.)

- Example 3: There will not be a monitor behaviour directly followed by monitor behaviour.

$$\text{not (Monitor} \xrightarrow{1, e = 1} \text{Monitor)}$$

This condition holds in both operationalisations so it is general over both.

- Example 4: There will not be a plan behaviour directly followed by a plan behaviour.

$$\text{not (Plan} \xrightarrow{1, e = 1} \text{Plan)}$$

This condition is violated in the Operationalisation 1 current solution and the Operationalisation 2 actual solutions. (Violated in part because the conception does not explicitly allow for planning without monitoring.)

5.4 INSPIRATIONAL INITIAL HCI ENGINEERING DESIGN PRINCIPLES FROM OPERATIONALISATION(S)

During operationalisation, potential initial HCI-EDPs are noted. Further potential initial such principles become clear during the investigation of other initial principles. These initial HCI-EDPs are also noted. These "hunches" deserve investigation. For example, it appears during operationalisation that "To achieve comfort with energy management systems in the home, prescribe 'late' control of the heating system by the user."

5.4.1 EXAMPLES

- Example 1: To achieve comfort with energy management systems in the home, prescribe "late" control of the heating system by the user.

This "hunch" is developed from Operationalisation 1, and "late" can be understood with respect to it. In Operationalisation 1, the control of the heating is moved to the end of the behaviours:

$$\text{Plan} : ...\text{Comfort} \xrightarrow[1,e\geq 1]{} \text{Plan} : ...\text{In house}$$
$$\Downarrow$$
$$\text{Plan} : ...\text{In house} \xrightarrow[1,e\geq 1]{} \text{Plan} : ...\text{Comfort}$$

A similar initial HCI-EDP in Operationalisation 2 can be looked for, that is, in its general form (or perhaps, to be true to the original "hunch," with P remaining as "Comfort"):

$$\text{Plan} : ...X \xrightarrow[1,e=1]{} \text{Plan} : ...P$$
$$\Downarrow$$
$$\text{Plan} : ...P \xrightarrow[1,e=1]{} \text{Plan} : ...X$$

Unfortunately, there is no obvious case. The latter can be identified as a *counter-principle*—an analysis not the case for particular operationalisations. Further operationalisations will enable more detailed generality (and probably further counter-principles).

- Example 2: More specific monitor→plan

Analysis of the first initial assumption corollary above leads to the hunch that there might be a more specific initial HCI-EDP based on the monitoring and planning parameters. For example:

$$\text{Monitor} : ...\text{Comfort} \xrightarrow[1,e\geq 1]{} \text{Plan} : ...\text{Comfort}$$

holds for Operationalisation 1 current and actual. The more general form:

$$\text{Monitor} : ...X \xrightarrow[1,e\geq 1]{} \text{Plan} : ...X$$

holds for both operationalisations.

- Example 3: Planning takes longer overall, is more effort overall, but provides the benefits.

If the planning effort equates to the structural and behavioural costs in planning, then the operationalisations can be compared. The actual time taken for planning could be used, but would be difficult to measure. So the event ticks are used, which is the same as the behavioural costs.

5.4 INSPIRATIONAL INITIAL HCI ENGINEERING DESIGN PRINCIPLES 75

Table 5.1: Planning: Operationalisations 1 and 2

Op1 Planning

Current structs		Actual structs		Difference	
Abstract	**Physical**	**Abstract**	**Physical**	**Abstract**	**Physical**
82	1	63	1	-19	0

Current behs		Actual behs		Difference	
Abstract		**Abstract**	**Physical**	**Abstract**	**Physical**
66	1	50	1	-16	0

Op2 Planning

Current structs		Actual structs		Difference	
Abstract	**Physical**	**Abstract**	**Physical**	**Abstract**	**Physical**
91	2	163	9	72	7

Current behs		Actual behs		Difference	
Abstract	**Physical**	**Abstract**	**Physical**	**Abstract**	**Physical**
214	5	1267	86	1053	81

The analysis clearly shows the differences between the two operationalisations. It is not generally the case that planning takes more effort overall (nor takes longer), giving another counter-principle.

- Example 4: Control effort is decreased and the benefits are provided.

Following from Example 3 above, perhaps it is the control effort that is more important in the prescription of solutions. A similar analysis to Example 3 can be performed by inspecting the control costs.

Table 5.2: Control: Operationalisations 1 and 2

Op1 Control

Current structs		Actual structs		Difference	
Abstract	**Physical**	**Abstract**	**Physical**	**Abstract**	**Physical**
51	25.9	38	28.63	-13	2.73

Current behs		Actual behs		Difference	
Abstract	**Physical**	**Abstract**	**Physical**	**Abstract**	**Physical**
59	29	34	10	-25	-19

Op2 Control

Current structs		Actual structs		Difference	
Abstract	**Physical**	**Abstract**	**Physical**	**Abstract**	**Physical**
346	98.5	390	108.34	44	9.84

Current behs		Actual behs		Difference	
Abstract	**Physical**	**Abstract**	**Physical**	**Abstract**	**Physical**
496	108	590	159	94	51

So, the generality is: "an increase in the physical structural costs."

The human costs can be separated from the computer costs:

Table 5.3: Control: Operationalisations 1 and 2—Human-only

Op1 Control (H only)

Current structs		Actual structs		Difference	
Abstract	**Physical**	**Abstract**	**Physical**	**Abstract**	**Physical**
35	7	24	6	-11	-1

Current behs		Actual behs		Difference	
Abstract	**Physical**	**Abstract**	**Physical**	**Abstract**	**Physical**
41	11	24	7	-17	-4

Op2 Control

Current structs		Actual structs		Difference	
Abstract	**Physical**	**Abstract**	**Physical**	**Abstract**	**Physical**
256	65	300	75	44	10

Current behs		Actual behs		Difference	
Abstract	**Physical**	**Abstract**	**Physical**	**Abstract**	**Physical**
382	70	488	125	106	55

There are no generalities (except counter-principle generality).

Table 5.4: Control: Operationalisations 1 and 2—Computer-only

Op1 Control (C only)

Current structs		Actual structs		Difference	
Abstract	Physical	Abstract	Physical	Abstract	Physical
16	18.9	14	22.63	-2	3.73

Current behs		Actual behs		Difference	
Abstract	Physical	Abstract	Physical	Abstract	Physical
18	18	10	3	-8	-15

Op2 Control (C only)

Current structs		Actual structs		Difference	
Abstract	Physical	Abstract	Physical	Abstract	Physical
90	33.48	90	33.26	0	-0.22

Current behs		Actual behs		Difference	
Abstract	Physical	Abstract	Physical	Abstract	Physical
114	38	102	34	-12	-4

The generality is: "a reduction in the computer control costs."

5.5 INITIAL HCI ENGINEERING DESIGN PRINCIPLES FROM GENERAL GUIDELINES

Guidelines are design knowledge and so might provide a basis for "interesting" initial HCI-EDPs. Particularly, if they are demonstrated to support best-practice.

General guidelines, such as "feedback" and "consistency," might be identified within the operationalisation(s). The difficulty would be in delimiting and defining the investigated guideline, because of their underdpecification. Any generality within the operationalisation(s) would support that delimiting and defining.

5.5.1 EXAMPLES

- Example 1: Feedback

Feedback might be further described as the provision of (relatively rapid) response by the computer after an input by the human. This can be represented as:

$$H : Input \xrightarrow[1, t \le 4s]{} C : Response$$

where $t \le 4s$ means that the time between the input and the response should be less than 4 (or any other best-practice number of) seconds.

In Operationalisation 1 control current:

$$X : \text{FxP} : \text{Press} \xrightarrow{1, t \leq 1s} C : O : \text{Turn on LED}$$

However, a more human factor (HF) view of feedback might be:

$$H : \text{Input} \xrightarrow{1, t \leq 4s} H : \text{Encode response}$$

Operationalisation 1 control current gives:

$$X : \text{FxP} : \text{Press} \xrightarrow{1, t \leq 1s} X : \text{Encode} : \text{LED}$$

The pattern also exists in Operationalisation 1 control actual (X:FxP:Press and X:Encode: LED off) and Operationalisation 2 control current and actual (H:FxP:Turn and H:Encode:Gas change in H+C:Change gas composite). Feedback is general, therefore, if "H:Input" is taken as H:FxP:Press and H:FxP:Turn, and "H:Encode response" as H:Encode:LED, H:Encode: LED off, and H:Encode:Gas. Feedback over both operationalisations control can be represented as:

$$H : \text{Input} \xrightarrow{1, t \leq 1s} H : \text{Encode response}$$

To support design, the generality is:

$$\Downarrow$$
$$H : \text{Input} \xrightarrow{1, t \leq 1s} H : \text{Encode response}$$

(The direction of design operator has been added to show that it is required in the design solution.)

- Example 2: Consistency

Consistency might simply be a different term for a subset of design knowledge. However, the possibility that consistency between the current and actual systems for goals with the two standard following behaviours is considered. This might be represented as:

$$H : \text{FP} : X \xrightarrow{1, e = 1} Y \xrightarrow{1, e = 1} Z$$
$$\Downarrow$$
$$H : \text{FP} : X \xrightarrow{1, e = 1} Y \xrightarrow{1, e = 1} Z$$

This representation could be generalised for any number of standard-following behaviours with an additional construct.

This representation of consistency is violated in both operationalisations. A likelihood factor to the direction of design could be added (consistency of this kind might not always be desirable), but it would be preferable to search for the stronger principle, maybe becoming more specific.

5.6 INITIAL HCI ENGINEERING DESIGN PRINCIPLES FROM MUSE GUIDELINES

MUSE supports the development, during existing system analysis, and expression, during design, of design knowledge that is specific to the system. In part, these "specific guidelines" are documented in the design recommendations and speculation columns of the MUSE tables. Initially, it is sensible to concentrate on specific guidelines that are used in the solution.

5.6.1 EXAMPLE

Example: *Avoid having X remember to turn the heating on or off, if possible*

The above specific guideline from Operationalisation 1 suggests possible generalisation with *Improvement in planning activities (particularly start time) should reduce flustering and therefore overheating* in Operationalisation 2, since improvement of planning might reduce later 'remembering'.

"Reduction in remembering" might be understood as a reduction in the costs of planning that leads, or intends to lead, directly to control. In Operationalisation 1, these are all of the planning behaviours except for X:FP:Plan. In Operationalisation 2 current planning, it is all of the planning behaviours except for X:FP:Plan. In the latter's actual planning, it is all of the planning except for X:ShSubPlan and X:FP:Plan.

Table 5.5: Planning: Operationalisations 1 and 2

Op1 Planning

Current behs		Actual behs		Difference	
Abstract		Abstract	Physical	Abstract	Physical
64	1	48	1	-16	0

Op2 Planning

Current behs		Actual behs		Difference	
Abstract	Physical	Abstract	Physical	Abstract	Physical
212	5	662	16	450	11

The generality does not hold.

5.7 INITIAL HCI ENGINEERING DESIGN PRINCIPLES FROM MUSE TASKS

The MUSE task diagrams also contain some of the above "specific guidelines." It is likely that the generalised products are more likely to be general "specific guidelines," termed "MUSE task guidelines." The generalised products are the (x) and the (y) products.

To support intra-initial HCI-EDPs, generalisation over the MUSE (x) and (y) products for the design for an operationalisation would be useful, to produce products that might be termed (xy) products. Included in these products would be selection constructs that indicate a task change from the (x) situation to the (y) situation. The selection entries could be marked with (x) or (y), to support the direction of design operator (⇓).

To support inter-initial HCI-EDPs, generalisation over these (xy) products would be useful, to produce products that might be termed (xy^n) products.

5.7.1 EXAMPLE

- Example: Cycle 1 Conceptual Task Model - CTM(xy)

Figure 5.1 shows an (xy) product between the General Task Model—GTM(x) and CTM(y)—products in the Cycle 1 MUSE application.

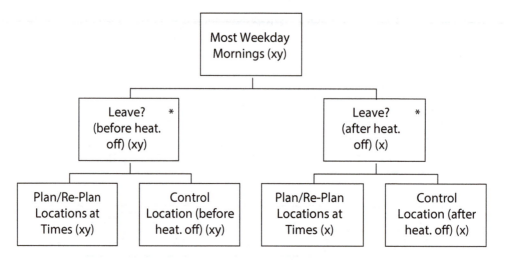

Figure 5.1: Cycle 1 Conceptual Task Model: CTM(xy) Product (following Stork, 1999).

The generality is unlikely to hold over Operationalisation 1 because it does not cover a current design scenario that includes leaving before the heating goes off. However, it shows an (xy) product. This example shows that generalisation over design scenarios with the same current artefact, user requirements, and artefact would be possible (analogous to MUSE TD analyses).

REVIEW

The chapter reports the instance-first strategy for acquiring initial HCI-EDPs in the domain of domestic energy planning and control. Operationalisation of the latter produces the identification of early, initial HCI-EDPs and inspirational initial such principles. The latter are also identified from general and MUSE guidelines and tasks. The concept of "counter-principle" is introduced, as an initial HCI-EDP not general across operationalisations.

5.8 PRACTICE ASSIGNMENT

5.8.1 GENERAL

Read § 5.1–7, concerning the initial HCI-EDPs.

- Check the initial HCI-EDPs informally for completeness and coherence, as required by the case study of domestic energy planning and control. Check the initial HCI-EDPs separately for principles identified during operationalisation (see § 5.2), inspirational principles from the operationalisation (see § 5.4), principles from general guidelines (see § 5.5), principles from MUSE guidelines (§ 5.6), and principles from MUSE tasks (see § 5.7).

- The aim of the assignment is for you to become sufficiently familiar with the initial HCI-EDPs to acquire them subsequently and, as appropriate, from a different domain of application.

Hints and Tips

Difficult to get started?

Re-read the assignment task carefully.

- Make written notes and in particular list the sections, while re-reading § 5.1–7.

- Think about how the sections might be applied to describe initial engineering principles from a novel domain of application.

- Re-attempt the assignment.

Test

List from memory as many of the sections as you can.

CHAPTER 6

Assessment and Discussion of Initial HCI Engineering Design Principles for Domestic Energy Planning and Control

SUMMARY

This chapter assesses the instance-first strategy for the acquisition of initial HCI-EDPs for the domain of domestic energy planning and control. The chapter comprises strategy assessment and discussion, also a proposal for MUSE for Research (MUSE/R).

6.1   STRATEGY ASSESSMENT AND DISCUSSION

This chapter assesses the research strategy for developing HCI-EDPs. The acquisition of such initial principles is discussed. The steps for acquiring HCI-EDPs from initial such principles are identified.

6.1.1   STRATEGY AND CONCEPTION CHANGES

The research strategy is initially described as bottom-up. However, it appears closer to top-down. First, the architecture conceptions and the planning and control conception directly influence the content of the initial HCI-EDPs. Second, the detailed strategy bases the identification of initial such principles on best practice of the time substantive design knowledge.

However, there remains a contrast with a top-down strategy. Stork and Long (1998) and Stork, Lambie, and Long (1998) describe a project that attempts the top-down strategy. They start with an informal statement of best-practice substantive design knowledge and then attempt to operationalise it as an initial HCI-EDP. Accordingly, Stork and Long propose that there might be a continuum of strategies between the bottom-up and the top-down—along the continuum of the expected initial generality. The present research strategy, then, is closer to the top-down strategy than originally anticipated. However, it can still be distinguished from the latter. Application of the alternative strategies outlined would support strategy selection.

Identifying initial expected generality in the strategy raises a concern for the conception of HCI-EDPs. If the specific design problem and its solution conception contain concepts that relate to principle acquisition, it might be questioned whether they are required for HCI engineering design practice. This might be the case, although a specific design problem and its solution conception might need to encompass alternative such general conceptions to match the partial design problem and solution operationalisations for the latter to be potentially applied. However, it seems more likely that it will be possible to operationalise partial design problems and solutions directly from the user requirements. Hill et al. (1995) implement a similar strategy that operationalises initial expected generality for a design model of the planning and control of multiple tasks.

6.1.2   STATUS OF INITIAL HCI ENGINEERING DESIGN PRINCIPLES

Initial HCI-EDPs have been acquired. The latter have the pre-requisites for acquiring potential guarantee. First, they are conceptualised according to a conception of the discipline of HCI (Long and Dowell, 1989). Second, they are operationalisations of design problem conceptions (Dowell and Long, 1989) based on the latter. Third, they are generalised over or within the two development cycles. Last, they are informally tested by successful evaluations of the two cycles.

The generality remains a concern for the initial HCI-EDPs. Particularly, two or fewer cycles are considered poor generality. This is prompted by the difficulty of selecting appropriate general cycle types. Further, the expression of the initial HCI-EDPs might not be appropriate for application to design practice. These concerns confirm that the initial HCI-EDPs should be considered "early."

6.1.3   STRATEGY ASSESSMENT

Early, initial HCI-EDPs are acquired. The strategy can be considered successful. Further cycles and validation are the next steps for assessing the strategy.

6.1.4   FURTHER RESEARCH

As concerns early, initial HCI-EDPs, the case study offers some examples. Further such principles could be identified from the research products.

Further development cycles are required to progress from early initial HCI-EDPs to initial such principles. To this end, more complex design scenarios need to be addressed.

Selection of the user requirements for development cycles is important. The user requirements here do not support potential generalisation as well as expected. In particular, the type of planning and control for each cycle artefact is different. The Cycle 1 artefact attempts to minimise re-planning, whereas the Cycle 2 artefact attempts to maximise pre-planning. An improved strategy might be to have a more rapid design phase before selection, perhaps encompassing the MUSE method (Lim and Long, 1994). Information Elicitation and Analysis Phase, analysing the current existing system to the Task Description TD (Current) and General Task Model GTM (Current) products. Also, the MUSE Design Synthesis Phase to the Conceptual Task Model CTM(y) product.

Validation of the initial HCI-EDPs involves re-expression and testing by application to a design scenario. Procedural HCI-EDPs are required for application. The guarantee of such principles, validated by application, needs to be based on first, the initial HCI-EDP guarantee, second, the operationalisations, and third, the (known) generality.

Testing is a challenge, however, since the effect of a particular HCI-EDP needs to be identified. The alternatives appear to be first, to control the designs to include or exclude the HCI-EDP application and, second, to "trace" the principle application and its contribution to effectiveness. Simulation may support this tracing. Metrification of the guarantee of HCI-EDPs could also be considered at this stage.

The research highlights a need for both procedural and tool support for the research strategy. Strategy products could be integrated with a method. A tool could also support, first the application of MUSE (a diagram editor was used), second, operationalisation, and third a detailed strategy–the identification of relationships. A tool could have supported the extension of "consistency" to e<=n and following behaviours. Last, is initial HCI-EDP validation.

6.1.5 FURTHER STRATEGY DISCUSSION

As concerns the current problem and solution, the specific design problem conception requires the current solution conception.

Operationalisation of conceptions requires several iterations. Particularly, the specific design problem domain operationalisation. Version 1 is derived from the current solution domain operationalisation. However, Version 2 is derived also from the specific design solution domain operationalisation, which is based on best-practice design. The latter might imply different goals from those initially expected.

As concerns best-practice and scientific knowledge, the strategy provides a means of incorporating best-practice and (applied) scientific knowledge into initial HCI-EDPs. Best-practice design knowledge is incorporated by best-practice design. Scientific knowledge is incorporated through the conceptions (for example, the cognitive architecture conception) and the operationalisation (for example, the formulae). However, note that neither best-practice nor applied scientific knowledge comes with a guarantee, concerning their incorporation into HCI-EDPs.

As concerns formal methods, the early, initial HCI-EDPs are formal in that they are operationalised to the level of metrics. The latter enable application of mathematical techniques. Expressed simply, the goal of formal methods is to base the software development process upon a workable set of mathematical techniques. However, one might argue against formal verification. Further, there are claims that the increasing complexity of computer programs will prevent them from being proven correct. The analogy here might be that HCI EDPs might be too complex to be formally applicable, so underlining the value of mathematics. However, the initial HCI-EDPs are not concerned with a scientifically correct description of the behaviour, only the utility of the initial

HCI-EDP. The requirement for formal specifications to be executable can be rejected. However, tool support may be required to apply HCI-EDPs. Further, the formalism may need to be altered to enable tool support.

The main concern for a formalism is whether it is useful and usable' (Stork, 1992). The known guarantee of HCI-EDPs ensures usefulness. Usability of HCI-EDPs relates to the requirement for that known guarantee (for example, safety critical systems may be worth significant effort).

6.2   MUSE FOR RESEARCH (MUSE/R)

A MUSE application involves the construction of products with a well-defined scope, process, and notation (§ 7.1). Similarly, the application of the research strategy involves the construction of products with a well-defined scope and notation. The scope and notation of the MUSE and the research products are related to each other to propose a version of MUSE to support research similar to the present.

6.2.1   SCOPE AND NOTATION

The research strategy has four main products: specific current design operationalisations; specific design problem operationalisations; specific design solution operationalisations; and initial HCI-EDPs. The scope of the products is expressed in the product name. The notations for the specific current design and specific design solution are: domain, structure, and behaviour diagrams; quality, structure changes; and behaviour formulae values. The notations for the problem and initial HCI-EDPs are logical/mathematical expressions.

MUSE has potential for supporting the research strategy because, first, most of the products are operationalisations of specific design scenarios, either systems being analysed or the system being designed. Second, the products have a well-defined and explicit scope.

The scope, processes, and notations of the MUSE products need to accommodate the research conceptions and to support the research strategy. MUSE has three phases: the Information Elicitation and Analysis Phase (IEA), the Conceptual Design Phase (CD), and the Detailed Design Phase (DD). The first IEA stages involve the analysis of existing systems—including operationalisations of the tasks and domains of discourse of the systems and generalisations of the tasks. For MUSE/R, these stages are re-scoped to operationalise: the specific design; the structures and behaviours of the existing worksystems; their structural and behavioural costs; their domain; the quality of their work; and appropriate generalisations of these operationalisations. Thus, the first IEA stages operationalise specific current designs according to the research conception.

The final IEA stages and the first CD stages of MUSE involve the specification of an HCI/human factors statement of the user requirements. For MUSE/R, these stages are re-scoped to operationalise the specific design problem. The final CD and DD stages of MUSE involve the specification of the interaction artefact and the documentation of the design rationale. For MUSE/R, these stages are re-scoped to operationalise the specific design solution and the previously acquired HCI/human factors best-practice knowledge applied to develop the specific design solution.

The notations for the re-scoping are those of the research strategy, similar or additional to the MUSE notations no longer employed.

6.2.2 PROCESS

The redefinition of the scope of MUSE/R suggests the need for process changes. Figure 6.1 shows the overview of MUSE/R.

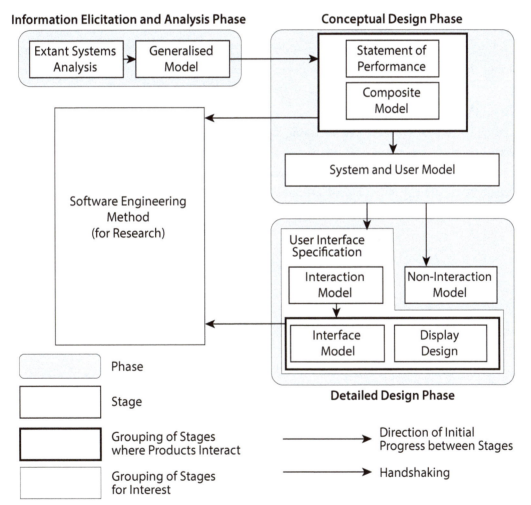

Figure 6.1: MUSE/R overview (following Stork, 1999).

6.2.3 SUPPORT FOR DESIGN

MUSE/R needs to support HCI practice at least as well as MUSE. The accommodation of the research conceptions suggests that the MUSE/R products are more complete and coherent (relative to the general design problem conception of HCI of Dowell and Long, 1989) than the MUSE products. The implication is for improved support for design.

6.2.4 FURTHER RESEARCH

Although an initial application of MUSE/R has been conducted, the scope, process, and notation of the products need to be defined further. Case studies are required to assess the support MUSE/R offers research and design. Further, the strong relationship between research and design in HCI itself needs to be investigated further (see also Long, 2021).

The evaluation component of the strategy could be incorporated into the MUSE/R method. Tool support for MUSE/R is essential. Automatic generation of prototype interfaces would be useful.

REVIEW

The initial HCI-EDPs are assessed as having the pre-requisites for acquiring potential guarantee. However, concerns are raised over their generality and their expression for application in design practice. They are, thus, considered to be early, initial HCI-EDPs. In spite of their early, initiual status, the research strategy is considered a success. A version of MUSE, termed MUSE for Research (MUSE/R), is proposed to support comparable HCI research.

6.3 PRACTICE ASSIGNMENT

6.3.1 GENERAL

Read § 6.1, concerning the strategy assessment and discussion.

- Check the sections informally for completeness and coherence, as required by the case study of domestic energy planning and control.

- The aim of the assignment is for you to become sufficiently familiar with the sections to apply them subsequently and as appropriate in the Practice Scenario 6.1.

Hints and Tips

Difficult to get started?

Re-read the assignment task carefully.

- Make written notes and in particular list the sections, while re-reading § 6.1.

6.3 PRACTICE ASSIGNMENT   89

- Think about how you might re-apply the sections to the application domain of domestic energy planning and control.

- Re-attempt the assignment.

Test

List from memory as many of the sections as you can.

Read § 6.2, concerning MUSE for Research (MUSE/R).

- Check the sections informally for completeness and coherence, as required by the case study of domestic energy planning and control.

- The aim of the practice assignment is for you to become sufficiently familiar with the sections to apply them subsequently and as appropriate in the Practice Scenario 6.2.

Read § 6.3, concerning conclusions.

- Check the sections informally for completeness and coherence, as required by the case study of domestic energy planning and control.

- The aim of the work assignment is for you to become sufficiently familiar with the sections to apply them subsequently and as appropriate in the Practice Scenario 6.3.

6.3.2   PRACTICE SCENARIOS

Practice Scenario 6.1: Formulate and Apply Your Own Strategy Assessment and Discussion

Formulate your own strategy assessment and discussion sections to replace those of § 6.1. Use the latter as a basis, for your formulation, along with examples from the HCI research literature.

- Apply your sections to the strategy assessment and discussion of the application domain of domestic energy planning and control (see § 6.1). The application can only be of the most general kind—that is at the level of the section. However, even application at this high level can provide practice for the researcher in assessing and discussing the strategy of their own work.

Practice Scenario 6.2: Formulate and Apply Your Own MUSE for Research (MUSE/R) Assessment and Discussion.

Formulate your own MUSE for Research (MUSE/R) sections to replace those of § 6.2.

- Use the latter as a basis, for your formulation, along with examples from the HCI research literature.

- Apply your sections to the MUSE for Research (MUSE/R) sections of the domestic energy planning and control application domain (see § 6.2).

Practice Scenario 6.3: Formulate and Apply Your Own Conclusions Assessment and Discussion.

Formulate your own conclusion sections to replace those of § 6.3. Use the latter as a basis, for your formulation, along with examples from the HCI research literature.

- Apply your sections to the conclusion sections of the domestic energy planning and control application domain.

CHAPTER 7

Introduction to Initial HCI Engineering Design Principles for Business-to-Consumer Electronic Commerce

SUMMARY

This chapter introduces HCI-EDPs for the domain of business-to-consumer electronic commerce. The introduction comprises: a conception of declarative and procedural HCI-EDPs; a class-first strategy for developing such principles; a method for operationalising the "class-first" strategy; and the identification of class design problems. This chapter constitutes the basis for the following chapter on the development of initial HCI-EDPs to support the more effective design of interactive business-to-consumer electronic commerce systems.

7.1 CONCEPTION OF HCI ENGINEERING DESIGN PRINCIPLES

To develop HCI-EDPs, Dowell and Long (1989) propose a conception of the HCI general design problem for an engineering discipline of HCI (Long and Dowell, 1989). The conception comprises work, the interactive worksystem (sic), and performance, expressed as task quality and worksystem costs.

However, the relationships between the concepts of the HCI general design problem and those of the general HCI-EDP remain to be specified. Likewise, for the general HCI-EDP relationship with the general design solution.

7.1.1 INTRODUCTION

The Dowell and Long (1989) conception of the HCI general design problem is summarised in § 7.1.2 to support the conception of the general HCI-EDP and the general design solution. Both conceptions comprise work, the interactive worksystem and performance, expressed as task quality and worksystem costs. The concepts of the general design problem are recruited to the conception of the general design solution. The former concepts constitute criteria for the success of the latter.

The conception of the general HCI-EDP includes: scope, comprising a class of users, a class of computers and a class of achievable performances; substantive component; methodological component; and guarantees. The relationships between the concepts of the general HCI-EDP and those of the general design problem and general design solution are specified.

7.1.2   CONCEPTION OF THE GENERAL HCI DESIGN PROBLEM

The Dowell and Long (1989) concepts are presented in *bold italics* on first appearance only, but apply throughout. A *design problem* expresses an inequality between *actual performance* (Pa) and *desired performance* (Pd) of some *interactive worksystem* (Pa ≠ Pd) with respect to some *domain*. A successful *design solution* specifies some interactive worksystem, which achieves desired performance (Pa = Pd) with respect to some domain. Worksystems comprise *users* and *computers*, both of which have *structures* supporting *behaviours*. Desired performance is expressed as work, achieved to a desired level of *task quality*, while incurring an acceptable level of *costs* to the worksystem. Work is expressed as *transformations* of the *attribute values* of *objects* in the domain of the worksystem. These domain transformations are achieved at a desired level of task quality (Tq), with an acceptable level of costs to the user (Uc) and to the computer (Cc). Attributes are features of domain objects, which afford transformation by the worksystem.

Worksystem goals are defined as a *product goal*, which is a transformation of object attribute values. Realisation of a product goal may involve the transformation of many attributes and their values. The latter are termed *task goals*. A product goal, then, may be re-expressed as a *task goal structure*, which specifies the order and relations between task goals, sufficient to achieve the product goal. As more than one task goal structure may achieve the same product goal, it is necessary to distinguish alternative task goal structures in terms of task quality. The latter describes the difference between the product goal and the actual transformation, specified by a task goal structure.

The difference supports evaluation of alternative structures of this type. The worksystem comprises one or more users interacting with one or more computers, each of which is characterised by structures, which support behaviours. Desired performance is thus effected by a particular class of user and computer structures, supporting behaviours, which achieve domain transformations, while incurring an acceptable level of costs. Knowledge of financial transactions supports transacting behaviours. The latter involve the transformation of object attributes and their values. For example, transferring ownership of goods from the vendor to the customer transforms the attribute "owner" from value "vendor" to value "customer" for domain object "item *x*."

7.1.3   CONCEPTION OF THE GENERAL HCI DESIGN SOLUTION

Concepts of Dowell and Long (1989), recruited to the conception of the general design solution and general HCI-EDP, are presented in *bold italics* on first appearance only, but apply throughout. A *design solution* contains the specification of a *worksystem* for which *actual performance* equals *desired performance* (Pa = Pd), as stated in the *design problem*. Worksystems, comprising *users* and *computers*, are conceptualised as *structures*, which support *behaviours*, which interact to perform *work* in a *domain*. Work is expressed as *transformations* of *object attribute values* to achieve *task goals*, which comprise a *task goal structure*. The *quality* with which the task goal structure achieves the *product goal*, specified in the general design problem, is expressed as Tq and the *costs* incurred by the users and computers are expressed as Uc and Cc, respectively.

7.1.4 CONCEPTION OF THE GENERAL HCI ENGINEERING DESIGN PRINCIPLE

The general HCI-EDP is complete if it identifies its applicability to the general HCI design problem. It is coherent if it is sufficient to prescribe design knowledge for specifying the general design solution. The ascription of performance guarantees is also required.

The conception of HCI-EDPs includes: scope, as a class of users, a class of computers, and a class of achievable performances; substantive component; methodological component; and performance guarantees. The conception is sufficiently coherent and complete to support the initial operationalisation, test, and generalisation of HCI-EDPs, and so their potential fitness-for-purpose to support design practice effectively.

This knowledge supports the operationalisation class HCI-EDPs. The latter specify the relationship between a class design problem and a corresponding class design solution. Classes support the representation of design knowledge at different levels of generality. These classes refer to the scope of class HCI-EDP class hierarchies, which include only class design problems for which a solution exists.

7.1.4.1 Scope of the General HCI Engineering Design Principle

Specifying criteria to identify *design problems*, to which an HCI-EDP may be applied, ensures that the latter is applied only to design problems for which it supports the specification of a *design solution*. Design problems contain not less than one, or more *users*, interacting with not less than one, or more *computers*, and some *desired performance*. The *scope* of the general HCI-EDP, then, comprises a *class of users* (U-class), a *class of computers* (C-class), and a *class of achievable performances* (P-class). If the user and computer, object of the design problem, are members of U-class and C-class, respectively, and the desired performance stated in the design problem is a member of P-class, then a design solution can be produced. The *actual performance* of the solution would equal the desired performance, as stated in the design problem.

The relationship between U-class, C-class, and P-class is developed by empirical testing of the implemented design solutions, produced by class HCI-EDP operationalisation. If the user, computer, or desired performance are outside the scope of the principle, then there is no guarantee that the design solution may be "specified then implemented." For transaction systems, the criteria for establishing U-class membership would establish the minimum structures and behaviours, required for some user, in conjunction with some member of C-class, to achieve a performance, which is a member of P-class. Such structures might include knowledge of financial transactions with card-based payment technologies. Supported behaviours might include matching goods descriptions to their shopping goals. The criteria for C-class membership might include structures such as "add to cart" button and supported behaviours, such as incrementing a list of items ordered by the user. P-class would specify the product goal, for example, support the exchange of resources for currency, which could be achieved by members of U-class and C-class, to a desired level of task quality, while incurring an acceptable level of costs.

7.1.4.2 Substantive and Methodological Components of the General HCI Engineering Design Principle

Dowell and Long (1989) argue that HCI-EDPs may be either declarative, that is substantive or procedural, that is methodological. Declarative HCI-EDPs 'prescribe the features and properties of artefacts, or systems that will, constitute an optimal design solution to a general design problem'. Procedural HCI-EDPs "prescribe the methods for solving a general design problem." Here, declarative and precedural knowledge are included together. The specification of declarative design knowledge and of the procedural knowledge, operationalised during its acquisition, are required aspects of HCI-EDP development.

HCI-EDPs contain declarative and procedural design knowledge, which may be applied to any design problem within their scope. The *declarative component* is characterised by the conceptualisation of user and computer *structures* and *behaviours*, comprising the *worksystem*, present in some instance of the class of users (U-class) or class of computers (C-class), respectively. The *procedural component* supports the conceptualisation of a *task goal structure*, comprising *task goals*, to be effected by the worksystem, which achieves the *product goal* stated in P-class. The product goal specifies the *work* to be effected in the domain by the worksystem, in terms of *object attribute value transformations*. The structures and behaviours of the declarative component are sufficient to achieve the task goal structure of the procedural component to an acceptable level of *task quality*, while incurring an acceptable level of *costs*. Task quality and worksystem costs are members of P-class. This sufficiency is supported by empirical testing of a class HCI-EDP, which indicates its fitness-for-purpose.

7.1.4.3 Summary of the General HCI Engineering Design Principle

A conception of HCI-EDPs, within which guarantees may be developed for HCI engineering design knowledge, is proposed. The conception comprises the following concepts and relationships. The summary supports application by other researchers.

For any design problem {user, computer, Pd} and an HCI-EDP {U-class, C-class, P-class, declarative component, procedural component}:

1. if user is a member of U-class and computer is a member of C-class, then user structures and behaviours and computer structures and behaviours, as stated in the declarative component, are present;

2. if user structures and behaviours and computer structures and behaviours as specified by the declarative component are present, then the task goal structure specified by the procedural component is achievable;

3. if the task goal structure specified in the procedural component is effected by a worksystem comprising the structures and behaviours specified in the declarative component, then the product goal will be achieved, task quality will be x, user costs will be y and computer costs will be z;

4. if task quality x, user costs y, and computer costs z are achieved, then Pa = Pd;

5. therefore, Pd is a member of P-class for a worksystem comprising instances of U-class and C-class.

Conception coherence derives from two relationships—that between the task goal structure and Tq for some product goal and that between the worksystem structures and behaviours, sufficient to achieve this task goal structure and Uc and Cc. These relationships are coherent, as performance is a function of the effectiveness with which some task goal structures are achieved by some worksystem structures and behaviours. The general HCI-EDP conception is complete as it comprises the concepts of the engineering discipline of HCI, which inform its development. The issue of fitness-for-purpose is addressed by operationalisation of the conception of the general HCI-EDP. The latter informs the development of class HCI-EDPs, which may then be tested and generalised.

7.1.4.4 Validation and Ascription of Guarantees to the General HCI Engineering Design Principle

Operationalisation of the general HCI-EDP, as class HCI-EDPs, supports empirical testing of the class-level design solutions prescribed. This testing establishes whether the general HCI-EDP is fit-for-purpose. That is, it supports the "specification then implementation" of a design solution, which achieves the desired level of performance, as stated in the design problem. The final stage of validation, which is generalisation, involves establishing the generality of the class HCI-EDP. These four stages of validation support the ascription of a guarantee. The latter is for a worksystem, which performs the task goal structure, specified in the procedural component of the engineering design problem, and achieves a level of Tq within the P-class, as stated in the HCI-EDP.

A second guarantee may be ascribed. The latter derives from the declarative component support of the specification of a worksystem, which exhibits the structures and behaviours sufficient to achieve the task goal structure. The latter is specified in the procedural component, while incurring a level of costs within the P-class, as stated in the HCI-EDP.

A third guarantee, that correct application of the HCI-EDP to a design problem within its scope supports the "specification then implementation" of a design solution, achieves Pd.

Ascription is assigned on the basis of the guarantees already identified and with further empirical testing. HCI-EDPs thus support the "specification then implementation" of a design solution, which achieves desired performance, if the design problem is within the scope of the HCI-EDP.

7.2 STRATEGY FOR DEVELOPING HCI ENGINEERING DESIGN PRINCIPLES

Sections 7.1.3–4 extend Dowell and Long's (1989) conception of the general design problem of HCI by specifying a conception of the general HCI-EDP and the general design solution. This section identifies two possible strategies for HCI-EDP development: "instance-first" and "class-first."

7.2.1 INTRODUCTION

The instance-first strategy was proposed and operationalised by Stork and Long (1994). The class-first strategy is presented and developed here. The class-first strategy involves identification of promising potential class design problems. The latter is based on the descriptive similarities between: specific design problems; specification of class design problems; specification of associated class design solutions; and derivation of specific design solutions from these class design solutions. These specifications enable actual performance of the class design solution to be established empirically.

7.2.2 INSTANCE-FIRST AND CLASS-FIRST STRATEGIES

Stork and Long (1998) applied the Dowell and Long (1989) conception of HCI to establish a basis for developing HCI-EDPs. They operationalise the general HCI design problem by instantiating its concepts and so rendering them observable and measurable. The domain is the planning and control of domestic energy. Operationalisation provides observable and measurable criteria against which to

7.2 STRATEGY FOR DEVELOPING HCI ENGINEERING DESIGN PRINCIPLES   97

assess performance. The design solutions of such specific design problems and the abstraction therefrom of prescriptive design knowledge would constitute an HCI-EDP. However, the operationalisation of a specific design problem does not, of itself, ensure that a class design problem, of which the specific design problem is an instance, exists. For this reason, the strategy is termed the instance-first strategy. It seeks to develop HCI-EDPs by specifying design knowledge for specific design problems, by means of specific design solutions, and then generalising across instances.

The approach is contrasted with the class-first strategy, proposed here. The latter requires the creation of class design problems and their associated class design solutions to establish the related design knowledge. This knowledge is considered promising for HCI-EDP development, as it is construed at a class level. The development of class design problems prior to operationalisation is rationalised, because it constrains the specific design problems operationalised to those which offer promise in supporting the identification of class-level knowledge.

7.2.3   HCI ENGINEERING DESIGN PRINCIPLES AS CLASS DESIGN KNOWLEDGE

HCI EDPs comprise conceptualised, operationalised, tested, and generalised design knowledge. The latter supports the diagnosis and prescription of a class of design solutions, which solve a class of design problems. HCI-EDPs are defined functionally, by reference to the class of design problems to which they may be applied. Thus, the specification of design problems and design solutions, at the class level, is a pre-requisite, and so necessary process of HCI-EDP development. To support the ultimate specification of such principles, there is first a need iteratively to identify class design problems and their associated class design solutions. Second, to extract both the commonalities between them, and the non-commonalities, which are exhibited only by the class design solution. The latter would constitute an initial HCI-EDP, which applies to all design problems within its scope. The characteristics of class design problems and class design solutions, and their interrelations, are specified here. Criteria for identification and selection of promising classes are also proposed.

7.2.4   CLASS DEVELOPMENT

The first stage of the class-first strategy for class HCI-EDP development involves identification of a specific design problem and a corresponding specific design solution. The second stage involves identifying further specific design problems, which require a similar Pd, which supports specification of P-class. The user(s) and computer(s), which are to achieve P-class, are then assessed for similarities. They are considered similar, if the user(s) and computer(s), specified in each specific design problem, comprise sufficient structures supporting sufficient behaviours to achieve P-class. If this sufficiency holds, these user(s) and computer(s) form U-class and C-class of the class design problem respectively. In practice, once P-class has been specified, developing class design problems involves identifying U-class and testing instances (members) of this class interacting with instances of C-class. These instances of U-class and C-class are then used to inform the development of a class design solution, which achieves P-class. The level of generality should be considered prior to development. Classes, which contain few instances and are low in the hierarchy, contain design knowledge, which is very specific. The costs of developing a class at a given level of generality needs to be balanced against the number of instances to which it may be applied successfully.

7.2.5 DEFINITION OF CLASSES

Each class of design problem contains instances, which inherit the characteristics of the parent class. These instances may be specific design problems or they may be (sub) classes. Such a hierarchy allows the categorisation of design problems at the appropriate level of generality. The hierarchy of design problems is mirrored by the hierarchy of design solutions, and by the hierarchy of principles, which effects the construction of the latter from the former. Such classes are abstractions, which are validated by their utility to support the definition of successful design solutions.

7.2.6 CONCEPTION OF CLASSES OF DESIGN PROBLEM

Design problems characterise the work that a technology achieves, in terms of domain transformations. The domain comprises objects, which have attributes, each of which has a specific state, or value. Work is characterised in terms of the transformation of object-attribute values, to be achieved by the worksystem. Task quality characterises the completeness with which the worksystem achieves the desired domain transformations.

The statement of desired performance (Pd) may be accompanied by a statement of performance for an existing worksystem. The statement expresses the actual performance (Pa). The design problem contains either a statement of ineffectiveness (Pa $\neq$ Pd), which makes explicit reference to the desired performance, or a statement of desired performance (Pd), if there is no existing worksystem. An acceptable solution has an actual performance, which equals that of desired performance (Pa = Pd), as the Pa of the solution equals (or exceeds) the Pd as stated in the design problem.

Any constraints on characteristics which the worksystem must or must not have, for example standard architectures and protocols, which must be adhered to, or user structures which are or are not available (for example, as the result of training, or the lack of it), are stated in the design problem. This statement allows issues, which may affect implementation of the solution, to be addressed by appropriate formulation of the problem to be solved.

7.2.7 CONCEPTION OF CLASSES OF DESIGN SOLUTION

The contents of any class of design solution are expressed in terms of worksystem structures and behaviours. A design solution specifies worksystem structures and their behaviours, which effect the domain transformations, as stated in the corresponding design problem to the level of effectiveness specified (Pd). The knowledge contained within this class may be considered as declarative state-

7.2 STRATEGY FOR DEVELOPING HCI ENGINEERING DESIGN PRINCIPLES   99

ments. The latter specify structures, which a worksystem must possess to achieve the performance requirements, detailed in the class of design problem. The corresponding procedural knowledge specifies the behaviours, which, if effected by the worksystem structures, achieves the product goal to the level of effectiveness stated in the class of design solution, that is, the task goal structure.

7.2.8 IDENTIFICATION OF PROMISING CLASSES

A class may be considered promising for development if, for some specific design problem, a corresponding specific design solution exists and there are other specific design solutions, which share features of the solved specific design problem. Once an initial class hypothesis has been formulated, the viability of the class may be assessed by examination of the work performed and the worksystem structures and behaviours sufficient to achieve Pd. If the performance achieved by the worksystem (Pa), specified in the specific design solution, is similar to the Pd of other specific design solutions (that is, Pa = Pd), then the specific design solutions show promise for class design principle development. The utility of any class specified is dependent on the following.

1. Fitness-for-purpose: knowledge which may be applied to any design problem within the scope of the class, such that the design solution specified exhibits Pa = Pd.

2. Class exclusivity: the class should have significantly different knowledge from other sub-classes in the same class.

3. Content: the class should contain instances, as an empty class does not qualify as a valid class for HCI-EDP development.

These requirements serve as criteria to assess promising potential class design problems. First, class design problems and their class design solutions are fit-for-purpose, if they allow appropriate categorisation of knowledge, such that it supports the production of specific design solutions, which satisfy specific design problems. Second, any instance in a set of sub-classes specified should vary in some way from its sister classes. This variation should indicate significant differences in design to ensure class descriptions are adequate for providing design support. The third criterion for viable classes is that any class should have some instances. This ensures that any class provides design support for some specific design problems.

The hierarchy of classes may be tested empirically by construction and examination of specific design solutions for specific design problems. The three criteria identified earlier are used to examine the initial viability of proposed classes with respect to achieving this function.

7.3 METHOD FOR OPERATIONALISING THE CLASS-FIRST STRATEGY

The method for operationalising the class first strategy comprises: introduction; class design problem and class design solution specification method; and identification of class design problems.

7.3.1 INTRODUCTION

The method for specifying class design problems and class design solutions produces a specification of a class design problem. The latter is established on the basis of descriptive similarities of specific design problems, identified from current best practice systems. The class design problem is assessed, before the class design solution is specified, using existing HCI design knowledge. The latter includes models and methods (see § 1.4.1.1) and principles, rules, and heuristics (see § 1.4.1.2). The class design solution is then decomposed as specific design solutions, to enable testing. The outcome of specific design solution testing supports the abstraction of actual performance for the class design solution. The relative performance of the class design problem and class design solution is specified using the costs matrix (Cummaford and Long, 1999), a representation for presenting worksystem performance, which enables systematic comparison of performances.

The method for the specification of HCI-EDPs involves identification of commonalities between a class design problem and its class design solution to form the scope of the HCI-EDP. Those aspects of the class design solution, which are not included in the class design problem-class design solution commonalities and the negation of those aspects of the class design problem, which are not included in the class design problem-class design solution commonalities, are then used to define the prescriptive component of the HCI-EDP. The latter's achievable performance is then defined as the actual performance achieved by the class design solution.

7.3.2 SPECIFICATION METHOD FOR CLASS DESIGN PROBLEM AND CLASS DESIGN SOLUTION

The following method specifies the class design problem and class design solution for Development Cycles 1 and 2. The stages are numbered in Figure 7.1.

7.3 METHOD FOR OPERATIONALISING THE CLASS-FIRST STRATEGY

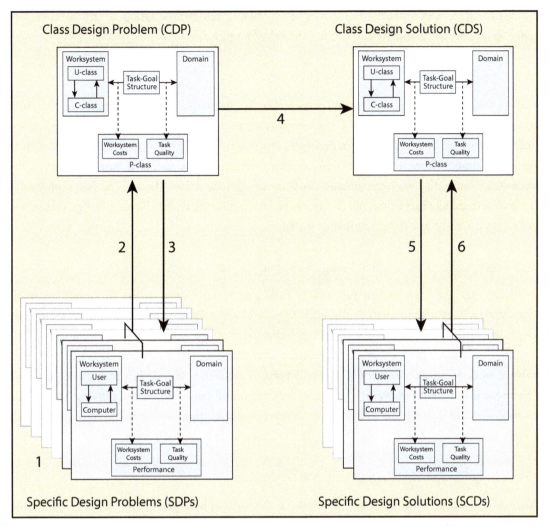

Figure 7.1: Method for class design problem and class design solution specification (following Cummaford, 2007).

7.3.2.1 Stage 1: Specify Specific Design Problems

A promising class of systems is identified on the basis of (informal) similarities in work performed. Examples of such systems are selected for testing. If these systems achieve a desired level of performance, they constitute specific design solutions. If Pa ≠ Pd, they constitute specific design problems. The specific design problem representations specify the domain objects, sufficient to characterise the work performed by the interactive worksystem, to achieve the product goal by operationalisation of the task goal structure. The interactive worksystem, comprising user and computer models, specifies the structures sufficient to support behaviours, to achieve the task goal structure. Performance, expressed as Tq and Uc and Cc, is established empirically.

7.3.2.2 Stage 2: Specify Class Design Problem

Following the specification of specific design problems, commonalities are abstracted to construct the class design problem. This abstraction comprises common aspects of the specific design problems, to provide an initial class design problem expression. The domain model is also abstracted, to express the product goal in terms of domain transformations. The task goal structure is then abstracted. The interactive worksystem model is likewise abstracted from the specific design problems. As class users are an abstraction, which cannot be tested directly, Pa for the class design problem is derived from the specific design problems tested.

7.3.2.3 Stage 3: Evaluate Class Design Problem

To ensure that the class design problem is sufficient to characterise specific design problem behaviours, it is evaluated analytically, with respect to the task goal structures for each specific design problem. The interactive worksystem behaviours of the class design problem are operationalised to achieve the task goal structure for each specific design problem. If the class design problem worksystem achieves the task goal structure, then the class design problem is retained. If there are insufficient behaviours, then specific design problem/worksystem differences must be re-synthesised. If they are too dissimilar to support synthesis, then a class design problem cannot be abstracted.

7.3.2.4 Stage 4: Specify Class Design Solution

Existing HCI design knowledge and methods, in the form of best-practice, are used to specify the class design solution. The design stage specifies a task design structure, which is achievable by the worksystem, while incurring acceptable costs and attaining desired Tq. While performance can only be established by testing specific design problems, instantiated from the class design solution, analytic methods can be used to establish the likely performance of the class design solution prior to testing.

7.3.2.5 Stage 5: Specify Specific Design Solutions

To evaluate the class design solution, it is necessary to re-express it as specific design solutions. As class users are an abstraction and so cannot be tested directly, specific design solutions must be designed to enable class design solution testing. It is necessary to instantiate more than one specific design solution to abstract class design solution commonalities in performance.

7.3.2.6 Stage 6: Evaluate Class Design Solution

The class design solution performance is abstracted from the performances attained by each of the specific design solutions. If Pa = Pd, then the class design solution is acceptable for the class design problem.

7.3.2.7 Costs Matrix

The need to evaluate the class design problem and class design solution systems in terms of their relative performance led to the development of the "costs matrix" (Cummaford and Long, 1999). The costs later provide a systematic means of representing the relationship between the work, how well it is performed by the worksystem, and the costs incurred by the user. Specification of these relationships facilitates the comparison of competing solutions to a design problem.

The costs matrix is constructed by listing the user abstract and physical behaviours on the y-axis and tasks on the x-axis. The cells of the matrix are filled by analysis of user data in the task goal structure. The cumulative abstract and physical costs are represented separately. No equivalence in cost is assumed between an abstract and a physical behaviour. The costs matrix also collates measurement of task quality, in terms of task completion rates and time taken to complete each task.

The costs matrix supports comparison of competing design solutions to a design problem in terms of their relative performance. However, such comparisons are only valid if the elements on both axes are the same for each analysed design solution and the means of evaluation is consistent between the systems analysed.

7.3.3 SPECIFICATION METHOD FOR HCI ENGINEERING DESIGN PRINCIPLES

The method for specification of HCI-EDPs requires the identification of commonalities between a class design problem and its class design solution, to form the scope of the HCI-EDP. Those aspects of the class design problem, which are not included in the class design problem—class design solution commonalities and the negation of those aspects of the class design problem, which are not included in the class design problem—class design solution commonalities, are then used to define the prescriptive component of the HCI-EDP. The latter's achievable performance is then defined as the actual performance of the class design solution.

An HCI engineering design problem comprises three components: a scope—supporting diagnosis; a specification—supporting prescription; and a class of achievable performances—supporting validation, the basis of guarantee, when applied. The HCI-EDP defines class design knowledge, recruited from class design solutions. The method for HCI-EDP construction is as follows.

7.3.3.1 Stage 1: Define Scope of HCI Engineering Design Principle

The HCI-EDP scope defines its boundary of applicability. The scope comprises a class of users and a class of computers, which interact to achieve a specified class of domain transformations within a specified class of domains. The scope is defined by generification of the commonalities between the class design problem and the class design solution. Defining the scope in this way ensures that sufficient components are in the class design problem to enable the class design solution components to be operationalised. An HCI-EDP is prescriptive class-level design knowledge, with a specified scope of application. The latter enables comparison with the scope of other related HCI-EDPs, to determine their relative generality.

7.3.3.2 Stage 2: Define Prescriptive Design Knowledge

The HCI-EDP prescriptive design knowledge is synthesised from the non-common aspects of the class design solution and the class design problem.

Stage 2.1: Identify Class Design Solution-only Components

The remaining components of the class design solution comprise the aspects of the class design solution, which, if operationalised for the HCI-EDP scope components, achieves the level of performance stated in the class design solution. The class design solution-only components are the foundation of the principle, offering a prescriptive specification of a task goal structure, user and computer representation structure states, supported user and computer behaviours (assumed to be commensurate with process structure activations), and achievable performance, as task quality and worksystem costs.

Stage 2.2: Identify Class Design Problem-Only Components

The method specified in Stage 2.1 is now used to identify those aspects of the class design problem, which are not represented in the class design problem-class design solution commonalities. While the class design problem-only components are not candidates for inclusion in the HCI-EDP, they may contribute to its specification by negation—that is, if the class design problem-only components are x, the HCI-EDP should specify not x.

Stage 2.3: Synthesise HCI Engineering Design Principle Prescriptive Component.

The class design solution-only and the negation of the class design problem-only components are now synthesised to construct the principle prescriptive component, comprising user and computer representation structure states and supported user and computer behaviours.

7.3.3.3 Stage 3. Define HCI Engineering Design Problem Achievable Performance.

The class design solution actual performance, as task quality and worksystem costs, is recruited from the class design solution to form the HCI-EDP class of achievable performances.

7.4 IDENTIFICATION OF CLASS DESIGN PROBLEMS

The identification of class design problems comprises: introduction; selection of potential class design problems; class of design problem (transaction systems); and transaction systems (specification of sub-classes).

7.4.1 INTRODUCTION

The class-first strategy, specified in § 7.2, involves identifying potential class design problems, which show promise for HCI-EDP development. A potential class design problem of transaction systems is identified. The latter expresses the need for the transaction system to "support the exchange of goods for currency," when certain conditions hold. This parent class contains instances (subclasses), each of which is also a class. Each subclass is characterised by performance, to be achieved by user(s) interacting with computers with respect to some domain. These subclasses are concerned with transaction systems for physical goods and for electronic goods—information and software, respectively. The general characteristics of each of these potential class design problems are inherited from the parent class design problem.

These classes are evaluated with respect to fitness-for-purpose, class exclusivity, and class content. The class design problem and class design solutions and initial HCI-EDP, for each of the two subclasses are developed. They are transaction systems for physical goods in Cycle 1 development and transaction systems for electronic goods information and software in Cycle 2 development.

7.4.2 SELECTION OF POTENTIAL CLASS DESIGN PROBLEM

To operationalise the class-first strategy, business-to-consumer electronic commerce transaction systems, hereafter transaction systems, are selected as a promising potential class of design problems. They are selected on the basis of informal similarities in the work carried out by existing systems, which can be characterised as "supporting the exchange of goods for currency between some customer and some vendor."

Once the initial class hypothesis has been formulated, the viability of the class can be assessed to identify commonalities by examination of the work to be performed and the worksystem structures and behaviours required to perform the work.

7.4.2.1 Work Communalities

Mercantile models, describing the process by which online transactions are effected, are identified from the literature and used to inform identification of commonalities in the work performed by the potential class of transaction systems and so assess class viability.

Mercantile Models

Kalakota and Whinston's (1996) model of the task elements of the online purchase of goods is shown in Figure 7.2. The model identifies seven transaction elements, grouped into three stages. Product selection and comparison elements are information gathering and decision-making activities, and so are not included in the identified commonalities. Post-purchase interactions, that is, returning goods, are not intrinsic to transaction systems at this level of abstraction. The return of goods is not always meaningful, for example, for transactions involving information access.

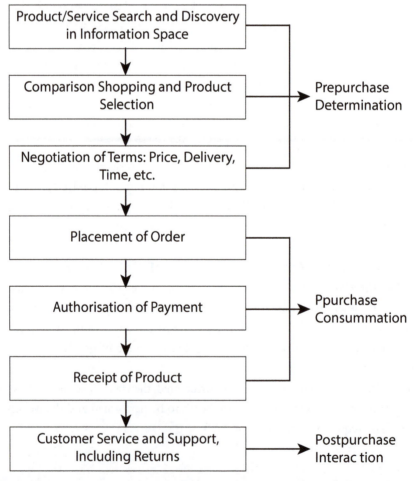

Figure 7.2: Kalakota and Whinston mercantile model (cited by Cummaford, 2007).

Elements of the model inform the specification of the following model of transactions.

Mercantile Model Transactions

The three stages now specified are proposed as the worksystem behaviours, that is, the task goal structure required to achieve the product goal identified, as "support the exchange of goods for currency between some customer and some vendor."

1. Negotiation involves the determination of the price, based on that of individual items ordered, any additional processing, for example, gift-wrapping and delivery charges. The latter vary as a function of the customer's location, the region or country to which the order is to be sent, that is, surface mail, air mail or courier delivery.

2. Agreement is reached when the customer and vendor agree on the goods to be supplied and the price and method of payment. Agreement involves the vendor stating the total price for the items specified by the customer, including the costs of any additional order processing and expedition, for the user to commit to purchasing.

3. Exchange comprises the transfer of payment, specified by the method agreed and the delivery of the goods.

Commonalities in worksystem structures and behaviours required to perform this task goal structure follow.

7.4.2.2 Worksystem Commonalities

The worksystem comprises a user and a computer—the "customer" and "vendor," respectively. To effect the task goal structure of the mercantile model, the customer must be in the vendor's domain of contract. The latter defines the physical locations to which a vendor may deliver the goods. The latter may be limited, for example, by local laws. The customer needs a delivery location both to determine whether they are within the vendor's domain of contract and to receive delivery of the goods.

The customer must have access to some payment technology, accepted by the vendor, for the exchange of goods for currency to occur. Current payment technologies used include credit cards and PayPal (an online money transfer service). The customer must also have sufficient funds to make the payment of the total price.

7.4.3 CLASS OF DESIGN PROBLEM FOR TRANSACTION SYSTEMS

The similarities in work and worksystem, identified earlier, are now used to define the product goal and domain model for the parent class of transaction system design problems.

7.4.3.1 Product goal

The work performed by transaction systems involves the exchange of resources between two (or more) transactants. The product goal for the class design problem is defined as: "for all customers, who fulfil the pre-purchase requirements and have the wherewithal and desire to purchase goods at stated total price":

1. transfer ownership rights of goods from vendor to customer and

2. transfer price of goods and any applicable surcharges (for example, sales tax, delivery costs) from customer account to vendor account.

The pre-purchase requirements are:

1. customer must be older than vendor's minimum age limit;

2. customer must be within vendor's domain of contract;

3. customer must have access to payment technology supported by the vendor;

4. vendor must offer for sale the items, which the customer wishes to purchase; and

5. customer must have sufficient funds to complete the transaction.

7.4.3.2 Domain Model

A design-oriented framework for the planning and control of multiple task work (PCMT), developed by Smith et al. (1997), is selected to inform development of the class design problem user, computer and domain models. The PCMT model employs concepts from the general HCI design problem (Dowell and Long, 1989), including the concepts of "domain" and "worksystem" and "user and computer structures and behaviours."

The domain must contain sufficient concepts to enable operationalisation of the product goal. It contains the objects "customer," "vendor," and "items" (that is, the goods), each of which has attributes, the values of which are transformed to achieve the product goal.

Specification of the domain model for the parent class of transaction system design problems, shown in Figure 7.3, indicates that the class offers promise for development. Two sub-classes identify and specify physical goods transaction systems and information transaction systems. Both class design problems are instances of the parent class of design problem for transaction systems and both inherit the characteristics of the class. The parent class design problem specified earlier is termed a general class.

7.4 IDENTIFICATION OF CLASS DESIGN PROBLEMS 109

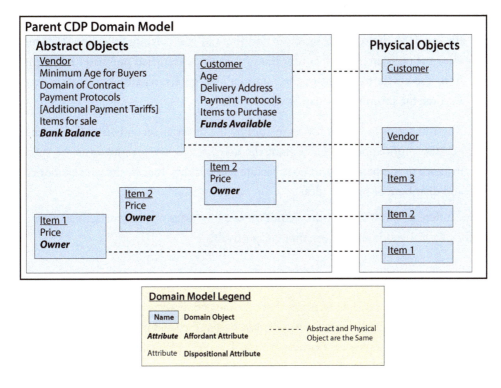

Figure 7.3: Parent class design problem for transaction systems: domain model (following Cummaford, 2007).

7.4.4 SPECIFICATION OF SUB-CLASSES FOR TRANSACTION SYSTEMS

Different characteristics by which the transactions of physical goods and information differ are identified by Hallam-Baker (1996). When goods are sold, the transaction is not completed until the goods are delivered. For the sale of information, payment is generally immediate and often inseparable from delivery of the information (for example, receiving a premium-rate Short Message Service or SMS) incurs automatic payment with no additional authorisation required). The different characteristics of physical goods and information are as follows.

1. Physical goods may be sold only once. Their potential value, then, is determined. The value of the goods is undetermined in the case of information, as a many potential sales are possible.

2. If the customer does not send payment or does so fraudulently, the seller of physical goods suffers an actual loss. The seller of information suffers a potential loss, if payment is not made, rather than an actual loss of resources.

3. The value of physical goods lies in the use of the physical artefact. Thus, returning the goods to the supplier in return for a refund is a meaningful action. Returning information is, in general, not meaningful, as the dissatisfied purchaser may benefit from the abstract content of the informational goods, regardless of future access. Further, the information may be easily copied.

4. Physical goods can generally be re-sold by the buyer as second-hand. The terms of sale for information generally grant the user limited usage rights over the specific instance of the product, and not absolute usage rights That is, the user cannot copy and distribute or sell the files.

These differences now inform the definition of sub-classes of the transaction system design problem, for physical goods and information, respectively.

7.4.4.1 Physical Goods as a Class of Design Problem for Transaction Systems

The domain model for the subclass of physical goods transaction systems design problems (CDP1—Class Design Problem 1) differs from the parent class in that the vendor has delivery cost tariffs as an attribute, and the user has a physical delivery address. The user's physical address determines whether the user is in the vendor's domain of contract, as required by the pre-purchase requirements (Figure 7.4).

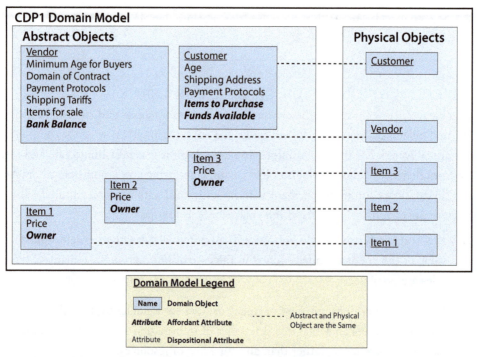

Figure 7.4: Class design problem 1: domain model (following Cummaford, 2007).

Examples of transaction systems in this class include (on-line) health supplement vendors, record shops, vintners, and motorcycle importers. Two systems are selected for development as specific design problems for Cycle 1 development. They are: Stash Tea (a U.S.-based retailer of tea, tea-making equipment, and associated products) and Herbs of Grace (a UK-based retailer of herbal remedies and aromatherapy products).

7.4.4.2 Information as a Class of Design Problem for Transaction Systems

The domain model for the subclass of electronic goods (information and software) transaction systems design problems (CDP2—Class Design Problem 2) differs from the parent class, in that the delivery cost tariff is not an attribute of the vendor and the customer has a mobile "phone number as the delivery address" (Figure 7.5). The "phone number is used to determine whether the customer is in the vendor's domain of contract, required by the pre-purchase agreement." The vendor offers subscriptions as well as products. The customer has an additional attribute "subscriptions to purchase" and the domain object "subscription" reflects this, with its attributes of price, subscription rights (that is, what the customer will receive in return for the price) and usage rights. Both single items and subscription rights have "usage rights" as an attribute, which specify the limited ways in which the customer can use the file, once downloaded. For example, they may use the information or software, but they cannot redistribute it to others.

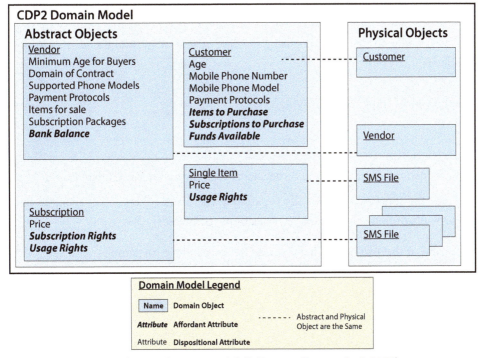

Figure 7.5: Class design problem 2: domain model (following Cummaford, 2007).

112 7. INTRODUCTION TO INITIAL HCI ENGINEERING DESIGN PRINCIPLES FOR B-TO-B

Examples of systems in this class include SMS news alert services and mobile phone software download services. Two transaction systems are selected for development as specific design problems for Cycle 2 development. They are Manchester United mobile news service, a UK-based provider of SMS alerts, containing news about Manchester United Football Club, and Jamster, a UK-based provider of ringtones, wallpapers and games for mobile phones.

7.4.4.3 Assessment of Subclasses

The classes defined earlier satisfy the requirements for successful class definitions.

1. The classes are fit-for-purpose, as they support the identification of transaction systems, in terms of their domain transformations.

2. The two subclasses have been demonstrated to have significant differences, in terms of their domain transformations.

3. The sub-classes contain instances and so offer promise for the eventual application of design knowledge, accrued during their development, to other design instances.

REVIEW

The chapter introduces HCI-EDPs for the application domain of business-to-consumer electronic commerce. The chapter comprises: a conception of HCI-EDPs; a class-first strategy for developing such principles; a method for operationalising the class-first strategy; and the identification of class design problems.

7.5 PRACTICE ASSIGNMENT

7.5.1 GENERAL

Read § 7.1, concerning the conception of declarative and procedural HCI-EDPs.

- Check the conception informally for completeness and coherence, as required by the case study of business-to-consumer electronic commerce.

- The aim of the assignment is for you to become sufficiently familiar with the conception to apply it subsequently and as appropriate to a different domain of application, as in Practice Scenarios 7.1–4.

Hints and Tips

Difficult to get started?

Re-read the assignment task carefully.

- Make written notes and in particular list the sections, while re-reading § 7.1.

- Think about how the sections might be applied to describe a novel domain of application.

- Re-attempt the assignment.

Test

List from memory as much of the conception as you can.

Read § 7.2, concerning the strategy for developing HCI-EDPs.

- Check the strategy informally for completeness and coherence, as required by the case study of business-to-consumer electronic commerce.

- The aim of the assignment is for you to become sufficiently familiar with the conception to apply it subsequently and as appropriate to a different domain of application, as in Practice Scenarios 7.1–4.

Read § 7.3, concerning the method for operationalising the "class-first" strategy.

- Check the method for operationalising the "class-first" strategy informally for completeness and coherence, as required by the case study of business-to-consumer electronic commerce.

- The aim of the assignment is for you to become sufficiently familiar with the method for operationalising the "class first" strategy to apply it subsequently and as appropriate to a different domain of application, as in Practice Scenarios 7.1–4.

Read § 7.4, concerning the identification of class design problems.

- Check the identification of class design problems informally for completeness and coherence, as required by the case study of business-to-consumer electronic commerce.

- The aim of the assignment is for you to become sufficiently familiar with the identification of class design problems to apply it subsequently and as appropriate to a different domain of application, as in Practice Scenarios 7.1–4.

7.5.2 PRACTICE SCENARIOS

Practice Scenario 7.1: Applying the Conception of HCI Engineering Design Principles to an Additional Domain of Application

Select a domain of application with which you are familiar or which is of interest to you or preferably both. The domain should be other than that of business-to-consumer electronic commerce.

- Apply the conception for business-to-consumer electronic commerce (see § 7.1) to the novel domain of application. The description can only be of the most general kind— that is at the level of the conception. However, even consideration at this high level can orient the researcher towards application of the conception to novel domains of application. The latter are as might be required subsequently by their own work. The practice scenario is intended to help bridge this gap.

Practice Scenario 7.2: Applying a Strategy for HCI Engineering Design Principles to an Additional Domain of Application

Select the same novel domain of application as for Practice Scenario 7.1.

- Complete as for Practice Scenario 7.1, but this time as concerns the application of a strategy for EDPs.

Practice Scenario 7.3: Applying a Method for Operationalising "Class-First" Strategy to an Additional Domain of Application

Select the same novel domain of application as for Practice Scenarios 7.1–2.

- Complete as for Practice Scenarios 7.1–2, but this time as concerns a method for operationalising the class first strategy.

Practice Scenario 7.4: Applying Identification of Class Design Problems to an Additional Domain of Application

Select the same novel domain of application as for Practice Scenarios 7.1–3.

- Complete as for Practice Scenarios 7.1–3, but this time as concerns the application of the identification of class design problems.

CHAPTER 8

Cycle 1 Development of Initial HCI Engineering Design Principles for Business-to-Consumer Electronic Commerce

SUMMARY

This chapter reports the Cycle 1 development of initial HCI-EDPs for the domain of business-to-consumer electronic commerce. This chapter comprises: Cycle 1 development and Cycle 1 class design problem/class design solution specification.

8.1 CYCLE 1 DEVELOPMENT

Section 7.4.2 identifies a potential class design problem for transaction systems, containing the two subclasses of physical goods and electronic information and software goods. Cycle 1 development addresses the class design problem and class design solution for the former. The method for identifying class design problems and their class design solutions is specified in § 7.3.2 and is used to organise this section.

Two specific design problems are selected to inform specification of the class design problem. They are physical goods online transaction systems for tea and related items and for health food supplements. The associated class design problem is then constructed from commonalities between the two specific design problems. The class design problem is then evaluated.

8.1.1 INTRODUCTION

A class design solution corresponding to the class design problem is specified and instantiated as specific design solutions. The latter correspond to the two specific design problems, identified earlier. The aim is to support testing. Class design solution performance is abstracted across specific performances of the constructed specific design solutions. The actual achieved performance equals the desired performance, specified in the class design problem. The class design solution is considered acceptable.

8.1.2 SELECTION OF SYSTEMS FOR SPECIFIC DESIGN PROBLEM AND SPECIFIC DESIGN SOLUTION DEVELOPMENT

Two electronic shops (e-shops) are selected. They exhibit similarities in the type of work supported, namely the sale of homogeneous physical goods. Both e-shops appear to have actual performances not equal to desired performance. They are promising for the specification of specific design problems. The e-shops selected were both existing commercial operations at the time. They are as follows.

- Specific Design Problem 1a: Stash Tea (see www.stashtea.com): a U.S.-based retailer offering a range of tea and tea-related products.

- Specific Design Problem 1b: Herbs of Grace (see www.herbsofgrace.co.uk): a UK-based retailer offering a range of herbal products, for example, essential oils, natural herbs, and herbal tablets.

The e-shops achieved similar product goals, but exhibited different behaviours.

8.1.3 TESTING PROCEDURE

Empirical testing is carried out following specification of the specific design solution systems, to support comparison of the relative workload incurred in using the specific design problem and specific design solution systems. The testing procedure follows to inform the later specification of specific design solutions.

8.1.3.1 Setup

To ensure the systems did not change during empirical testing, screenshots were used to create prototypes (offline simulations) of each e-shop. Prototypes ensured participants interacted with the same e-shop design. Also, that the analytical data in the associated costs matrix are based on the same system as the empirical data.

The specific design problem prototypes were in PowerPoint from screenshots of the live websites. Each presentation contained sufficient screens for completion tasks specified in the specific design problem product goal. Additional screens anticipated participants selecting links on incorrect task paths. Shopping task-related data were also included, such as descriptions and prices of the goods.

The specific design solution prototypes were also in PowerPoint. Graphical elements from the live sites were included, as well as product descriptions with prices and goods subtotals, to provide correct data to support task completion.

8.1.3.2 Participants

Six participants meeting the specific design problem criteria were recruited for each specific design problem/solution—three male and three female participants. Both specific design problems had the following pre-purchase requirements. The customer must be older than the vendor's minimum age limit (for specific goods) and within vendor's domain of contract. The customer must have access to payment technology supported by the vendor, and sufficient funds to complete the transaction. The vendor must offer for sale the items, which the customer wishes to purchase.

All participants resided in the UK and so were in the domain of contract. They were all over 18 and had used a credit card to purchase goods online. So, all met the pre-purchase requirements.

The Specific Design Problem 1a and Specific Design Problem 1b user models both specify that the user knows about shopping, payment, value for money, and personal wherewithal. Since participants had already used an online shop, they possessed the abstract structures specified in the specific design problem and specific design solution user models. None of the participants had used the tested shops.

The order of design problem and design solution presentation was balanced across participants to counter any learning effects. Participants were shown only one design problem/design solution pair (that is, 1a or 1b) and not both pairs, also to avoid learning effects.

8.1.3.3 Procedure

Testing was carried out at participants' own workstations, or at a dedicated PC. Each test was recorded on video. Each participant was asked to complete the shopping tasks specified for that specific design problem/solution pair, using the prototype systems.

Mouse movement and link selection was indicated by the participant pointing at the relevant link on-screen. The tester then moved to the appropriate screen in the simulation. Keystrokes were simulated by the participant typing the relevant keystrokes on a keyboard, unconnected to the computer. While the timing of the simulated actions may not have corresponded exactly to use of a live web application, it was sufficient for comparison of the specific design problem and specific design solution systems.

Task completion and time-to-complete were recorded for each task by the tester, using a data sheet. Time to complete ended, when the participant gave the correct answer or signalled task completion. Incorrect answers or uncompleted tasks were recorded as failed (coded as "0" in the costs matrices).

After attempting the shopping tasks on the first prototype, the participant completed a Likert scale as a subjective assessment of workload. The scale categories were: much too high; too high; acceptable; quite low; and very low.

The participant completed the same scale after using the second system, to enable comparison. Participants then completed an additional scale, comprising categories of: 1 was much higher than 2; 1 was higher than 2; about the same; 2 was higher than 1; and 2 was much higher than 1.

The latter scale was for comparing the relative workload, involved in using the two prototypes. These data were collected to evaluate the analytic measurements of workload (user costs) in the costs matrices (see § 7.4.2.1).

8.1.3.4 Testing Tasks

Seven shopping tasks were specified for each of the systems tested. The tasks are based on the Mercantile Model and also designed to include a range of behaviours, encountered during typical shopping activities. Tasks 1-3 involve product selection, and are designed to include multiple means of accessing products. That is: product listings on the Homepage in Task 1; products in a page of search results in Task 2; and products accessed via category listing pages in Task 3. Tasks 4–6 then correspond to the negotiation phase. The tasks involve the participant finding the total price for the order, before amending the order. This is to reduce the total price, before confirming the new total price. Task 7 corresponds to the agreement and exchange phases of the Mercantile Model (see § 7.4.2.1).

Similar tasks were specified for specific Design Problem/Solution 1a and Specific Design Problem/Solution 1b, the only differences being the actual products to purchase, which are specific for each shop. The seven tasks are:

1. order Item 1 from Homepage;

2. search for Item 2 and order;

3. go to the relevant category listing page and order 2 units of Item 3;

4. find total price;

5. delete 1 unit of Item 3;

6. find total price; and

7. buy goods.

8.1.3.5 Calculation of User Costs

The task goal structure contains descriptions of the user and computer tasks, required to complete each task. Also contained are the user abstract and physical behaviours. The latter are identified, using the criteria shown in Table 8.1.

8.1 CYCLE 1 DEVELOPMENT   119

Table 8.1: Criteria for diagnosing user abstract behaviours

User Behaviours	Criteria
Encoding	User reads one page of information. If scrolling is required, another "encode" behaviour is diagnosed for each additional screen of information. If the user registers that some information on the page has been updated in response to their recent action, this is not counted as an encode behaviour
Planning	Change state of a user model abstract representation (that is, transforms current plan for shopping)
Controlling	Determine next action to achieve current plan for shopping
Executing	"Transcode" abstract behaviour into physical behaviours

The costs matrices show the user costs incurred for tasks. The costs are calculated normatively to express workload for ideal, error-free task completion. This calculation enables comparison of the efficiency of the specific design problem and specific design solution systems in terms of their ideal operation. The costs matrices also include the empirical performance data from testing to determine their relative effectiveness.

8.1.4   SPECIFY SPECIFIC DESIGN PROBLEMS

The results from the testing of Specific Design Problems 1a and 1b are presented next to exemplify the behaviours resulting in actual performance not equalling desired performance.

8.1.4.1 Specific Design Problem1a

Desired performance (Pd) for Specific Design Problem1a required 100% task completion. Participants were asked to complete the following tasks.

- Task 1: Order 1 unit of cookies from Homepage.

- Task 2: Search for Brown Betty Teapot and order 1 unit.

- Task 3: Order 2 units of Black Tea.

- Task 4: Find the total price for your order (that is, what you will actually pay, including delivery).

- Task 5: Delete 1 unit of Black Tea from your order.

- Task 6: Find the new total price for your order (that is, what you will actually pay, including delivery).

- Task 7: Complete your purchase (that is, pay for your order).

120 8. CYCLE 1 DEVELOPMENT OF INITIAL HCI ENGINEERING DESIGN PRINCIPLES: B-TO-B

The costs matrix in Table 8.2 shows that no participants successfully completed Task 4 or Task 6 and, as such, Pa did not equal Pd. The order of presentation is shown to the right of the costs matrix.

Task 4 delivery costs were difficult to find. Participants gave up on Task 6, as they had not previously found the delivery costs.

Table 8.2: Specific Design Problem 1a cost matrix

		T1	T2	T3	T4	T5	T6	T7	totals	Order
Abstract behaviours	Plan	3	3	3	2	3	2	3	19	
	Control	4	6	7	7	3	6	9	42	
	Encode	3	4	4	7	2	7	4	31	
	Execute	2	3	4	3	2	3	8	25	
	Total	**12**	**16**	**18**	**19**	**10**	**18**	**24**	**117**	
Physical behaviours	Search	3	4	3	7	2	7	4	30	
	Click	2	8	3	3	2	3	25	46	
	Keystroke	0	8	2	0	2	0	92	104	
	Total	**5**	**20**	**8**	**10**	**6**	**10**	**121**	**180**	
Task completion	U1	1	1	1	0	1	0	1	5	DS / DP
	U2	1	1	1	0	1	0	1	5	DS / DP
	U3	1	1	1	0	1	0	1	5	DS / DP
	U4	1	1	1	0	1	0	1	5	DP / DS
	U5	1	1	1	0	1	0	1	5	DP / DS
	U6	1	1	1	0	1	0	1	5	DP / DS
	Total	**6**	**6**	**6**	**0**	**6**	**0**	**6**	**30**	
Time to complete	U1	14	27	2	183	7	39	83	355	
	U2	19	31	88	80	16	25	69	328	
	U3	42	22	29	193	31	26	72	415	
	U4	28	18	59	72	31	23	66	297	
	U5	45	32	36	153	4	44	65	379	
	U6	42	23	38	145	7	54	61	370	
	Total	190	153	252	826	96	211	416	2144	
	Total / 6	**31.7**	**25.5**	**42**	**137.7**	**16**	**35.2**	**69.33**	**357.333**	

The analytically calculated abstract and physical behaviours for Task 4 and Task 6 reflect the user costs incurred in following the correct task path, although no participants actually completed the tasks.

8.1.4.2 Specific Design Problem1b

Participants completed the following tasks, with the specific design problem and specific design solution systems:

- Task 1: Order 1 unit of St. John's Wort 100 Vcaps from Homepage.

- Task 2: Search for Gingko Biloba 100 Vcaps and order 1 unit.

- Task 3: Order 2 units of Tea Tree Essential Oil.

- Task 4: Find the total price for your order (that is, what you will actually pay).

- Task 5: Delete 1 unit of Tea Tree Essential Oil from your order.

- Task 6: Find the new total price for your order.

- Task 7: Complete your purchase (that is, pay for your order).

The six participants did not take part in the Specific Design Problem/Solution1a testing to avoid learning effects. However, Task 4 and Task 6 (both "find total price including shipping") again had less than 100% completion. The Specific Design Problem1b costs matrix is shown in Table 8.3. The errors shown for Task 4 and Task 7 are due to participants not calculating the correct task total amount.

Table 8.3: Specific Design Problem 1b cost matrix

		T1	T2	T3	T4	T5	T6	T7	Totals	Ordering
Abstract behaviours	Plan	3	3	3	2	3	2	5	21	
	Control	3	6	7	7	3	5	8	39	
	Encode	2	4	4	4	2	4	6	26	
	Execute	1	3	4	6	2	3	5	24	
	Total	**9**	**16**	**18**	**19**	**10**	**14**	**24**	**110**	
Physical behaviours	Search	2	4	3	4	2	4	5	24	
	Click	1	4	3	18	2	3	10	41	
	Keystroke	0	12	2	76	2	0	16	108	
	Total	**3**	**20**	**8**	**98**	**6**	**7**	**31**	**173**	
Task completion	U1	1	1	1	0	1	1	1	6	DS then DP
	U2	1	1	1	1	1	1	1	7	DP then DS
	U3	1	1	1	1	1	1	1	7	DP then DS
	U4	1	1	1	1	1	1	1	7	DS then DP
	U5	1	1	1	0	1	0	1	5	DP then DS
	U6	1	1	1	1	1	0	1	6	DS then DP
	Total	**6**	**6**	**6**	**4**	**6**	**4**	**6**	**38**	
Time to complete	U1	8	41	58	70	32	38	18	265	
	U2	6	114	64	107	20	37	44	392	
	U3	5	29	41	72	18	37	35	237	
	U4	9	61	53	163	12	26	51	375	
	U5	5	21	19	20	17	5	81	168	
	U6	4	22	34	112	19	36	57	284	
	Total	37	288	269	544	118	179	286	1721	
	Total / 6	**6.17**	**48**	**44.8**	**90.7**	**19.7**	**29.8**	**47.7**	**286.8**	

To complete Task 4, participants had to complete the first stage of checkout by entering their delivery address and then click through an order confirmation screen, before they could view shipping cost.

8.1.5 SPECIFY CLASS DESIGN PROBLEM

Following specific design problem specification, commonalities are abstracted to construct the class design problem domain model, product goal, task-goal structure, and interactive worksystem models (user model and computer model).

8.1.6 EVALUATE CLASS DESIGN PROBLEM

The evaluated e-shops exhibited different behaviours, resulting in a similar achieved product goal, but with different performances. However, the aspects of the work, resulting in high workload, were similar and included in the class design problem. The latter user model and domain model were operationalised analytically, to check the task goal structure in specific design problems could be achieved. The user model contains sufficient behaviours to achieve the task goal structure in each of the specific design problems tested. The class design problem, then, is retained.

8.1.7 SPECIFY CLASS DESIGN SOLUTION

The class design solution comprises: a product goal; domain model; user model; computer model, and task-goal structure. The class design problem product goal, domain model and user model are carried forward to the class design solution. However, the computer model (including the physical structures embodied as screens), and the task goal structure are designed during operationalisation. A combination of task goal structure re-engineering, and best-practice HCI design techniques are used, to construct a class design solution, achieving Pd.

Initial design involves re-engineering the class design solution task goal structure. First, as many behaviours as possible are allocated to the computer. Behaviours are then removed to leave the minimum behaviours required to effect the user and computer structure activations. The latter are necessary to achieve the domain transformations, specified in the product goal.

The computer structures sufficient to support the user and computer behaviours in the task goal structures are then identified, and grouped into screens.

8.1.8 SPECIFY SPECIFIC DESIGN SOLUTIONS

The class design solution is instantiated as specific design solutions, corresponding to the specific design problems identified. The former are then used to specify prototypes for empirical testing. The results from both Specific Design Solutions 1a and 1b indicate a 100% task completion rate, which is desired task quality performance.

A reduction in user costs, relative to those of the specific design problem systems, is desirable. The costs matrices for Specific Design Solutions 1a and 1b indicate that user costs are consistently lower, when interacting with the design solution rather than the design problem prototypes. Specific Design Solutions 1a and 1b are, thus, considered design solutions.

Participants' subjective ratings of the workload indicate that the design solutions systems generally require lower workload than the design problem systems.

8.1.9 EVALUATE CLASS DESIGN SOLUTION

The class design solution performance is abstracted from the performances attained by each of the specific design problems. The costs matrix contains the mean value for each element from the specific Design Solutions 1a and 1b costs matrices. Since the mean task completion rate across individual participants is not a valid measure, the empirical components (task completion and time to complete) are calculated using mean values across all participants for each system. Costs matrices are shown in Tables 8.4 and 8.5. They enable comparison of the two systems.

Table 8.4: Class Design Problem 1 costs matrix

		T1	T2	T3	T4	T5	T6	T7	Totals
Abstract behaviours	Plan	3	3	3	2	3	2	4	20
	Control	3.5	6	7	7	3	5.5	8.5	40.5
	Encode	2.5	4	4	5.5	2	5.5	5	28.5
	Execute	1.5	3	4	4.5	2	3	6.5	24.5
	Total	**10.5**	**16**	**18**	**19**	**10**	**16**	**24**	**113.5**
Physical behaviours	Search	2.5	4	3	5.5	2	5.5	4.5	27
	Click	1.5	6	3	10.5	2	3	17.5	43.5
	Keystroke	0	10	2	38	2	0	54	106
	Total	**4**	**20**	**8**	**54**	**6**	**8.5**	**76**	**176.5**
Task completion	**Percentage**	**100.0%**	**100.0%**	**100.0%**	**33.3%**	**100.0%**	**33.3%**	**100.0%**	**81.0%**
Time to complete (secs)	**Average**	**18.917**	**36.75**	**43.417**	**114.17**	**17.833**	**32.5**	**58.5**	**322.08**

Table 8.5: Class Design Solution 1 costs matrix

		T1	T2	T3	T4	T5	T6	T7	Totals
Abstract behaviours	Plan	5	3	3	2	4	2	3	22
	Control	4	5	5	0	3	0	12	29
	Encode	4	2	3	1	3	1	5	19
	Execute	2	2	2	0	2	0	10	18
	Total	**15**	**12**	**13**	**3**	**12**	**3**	**30**	**88**
Physical behaviours	Search	3	2	3	0	3	0	5	16
	Click	3	3	3	0	2	0	27	38
	Keystroke	0	10	0	0	0	0	107	115
	Total	**6**	**13**	**6**	**0**	**5**	**0**	**139**	**169**
Task completion	**Percentage**	**100.0%**	**100.0%**	**100.0%**	**100.0%**	**100.0%**	**100.0%**	**100.0%**	**100.0%**
Time to complete (secs)	**Average**	**10.5**	**16.333**	**20.083**	**19.583**	**13.583**	**5.0833**	**57.417**	**142.583333**

Participants achieved 100% task completion, when interacting with the class design solution, while incurring reduced costs. The former, then, is acceptable, as its success criteria are met.

The participant workload ratings and workload comparisons indicate that less workload was required for the specific design tasks. The latter systems, and by extension the class design solution, are then acceptable, fulfilling the requirement for lower costs.

The class design solution is retained.

8.2 CYCLE 1 CLASS DESIGN PROBLEM/CLASS DESIGN SOLUTION SPECIFICATION

8.2.1 INTRODUCTION

The class design problem and class design solutions are presented in terms of: work, as a product goal and domain model; an interactive worksystem, as a user model and computer model, and performance, as task quality and worksystem costs, as expressed by the costs matrix. These models inform the specification of an initial HCI-EDP.

The models, specified by the operationalisation of the method, are presented. The stages of the method are followed. For each specific design problem/solution and class design problem/solution model, its components are presented in bold on first exposition to enable mapping to the conception of class design problems and class design solutions.

8.2.2 STAGE 1: SPECIFY SPECIFIC DESIGN PROBLEMS

Two design problems are specified.

Specific Design Problem 1a

Domain and Product Goal

Specific Design Problem 1a is a transaction system supporting the exchange of tea plus related goods and equipment for currency. The task involves purchasing four specific items. One of the items has to be removed from the order to satisfy the user's financial constraint. The task was: T1: Order 1 unit of cookies from Homepage; T2: Search for Brown Betty Teapot and order 1 unit; T3: Order 2 units of Black Tea; T4: Find the total price for your order (that is, what you will actually pay, including shipping); T5: Delete 1 unit of Black Tea from your order; T6: Find the new total price for your order (that is, what you will actually pay, including shipping); and T7: Complete your purchase (that is, pay for your order).

The **domain model** comprises the user, computer (e-shop), and the goods to be purchased—see Figure 8.1. The domain model contains **abstract objects**, which are embodied in **physical objects**.

The **affordant domain attributes** (shown in bold italics in the model) are changed by the worksystem, in order to achieve the product goal. The **dispositional domain attributes** must have the values of the product goal for the work to be affected.

8.2 CYCLE 1 CLASS DESIGN PROBLEM/CLASS DESIGN SOLUTION SPECIFICATION

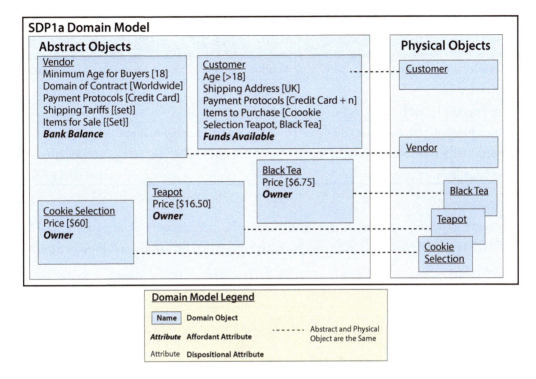

Figure 8.1: Specific Design Problem 1a Domain Model (following Cummaford, 2007)

The **Product Goal** specifies the required values for the dispositional domain object attributes, that is, the pre-purchase requirements, and a specification of the affordant domain attribute value transformations that comprise the work, that is, their start states and end states; see Tables 8.6 and 8.7.

Table 8.6: Specific Design Problem 1a product goal: dispositional object attribute value requirements	
Prepurchase Requirements	**Achieved?**
Customer: age [>18] >= Vendor: minimum age for buyers [18]	Yes
Customer: shipping address [UK address] is a member of vendor: domain of contract [global]	Yes
Customer: Payment protocols [{Credit card, Switch, PayPal}] and Vendor: Payment protocols [{Credit card}] must contain a common item	Yes
Customer: items to purchase [{Cookies selection, teapot, black tea}] must be members of Vendor: items for sale [{set, includes Cookies selection, teapot, black tea}]	Yes
Customer: funds available [$120] must be >= sum of price for every item in user: items to purchase, plus shipping costs as determined using vendor: shipping tariffs [$113.80]	Yes

Table 8.7: Specific Design Problem 1a product goal: Affordant object attribute value transformations

Domain Object: Attribute [value]	Start State	End State
Cookies selection: owner [value]	Vendor	Customer
Teapot: owner [value]	Vendor	Customer
Black tea: owner [value]	Vendor	Customer
Customer: items to purchase [{set}]	Cookies selection, teapot, black tea	
Customer: funds available [amount]		$6.20
Vendor: bank balance [amount]		$1,113.80

Worksystem

The worksystem comprises a user, termed "SDP1a-U," who interacts with a computer, termed "SDP1a-C." SDP1a-U and SDP1a-C both comprise Representation Structures and Process Structures.

User model

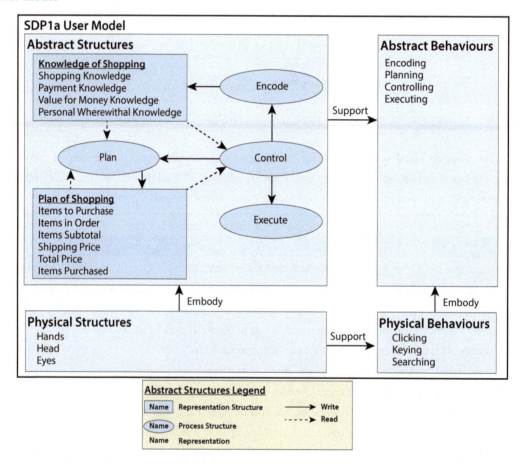

Figure 8.2: Specific Design Problem 1a User Model (following, § 7.4.2.1).

8.2 CYCLE 1 CLASS DESIGN PROBLEM/CLASS DESIGN SOLUTION SPECIFICATION   127

The user model (see Figure 8.2) comprises two types of abstract structure. **Representation Structures** (shown in boxes in the model) have particular states, for example, "items ordered," which are transformed by **process structures** (shown in ovals in the model, for example, "encode"). The representation structure states for each stage of the work are detailed in the representation structure states matrix; see Table 8.8. Process structure activation of representation structures is assumed to support user abstract behaviours. The user abstract behaviours, exhibited during the work, are specified in the task goal structure, described later.

Table 8.8: Specific Design Problem 1a user model representation structure states matrix

	Start	After T1	After T2	After T3	After T4	After T5	After T6	After T7
Abstract Structures								
Shopping knowledge	Starting state	Plus T1 increment	Plus T2 increment	Plus T3 increment	Plus T4 increment	Plus T5 increment	Plus T6 increment	Plus T7 increment
Payment knowledge	Starting state							Plus T7 increment
Value for money knowledge	Starting state	Plus T1 increment	Plus T2 increment	Plus T3 increment	Plus T4 increment	Plus T5 increment	Plus T6 increment	Plus T7 increment
Personal wherewithal knowledge	Starting state							Plus T7 increment
Plan for shopping								
Items to purchase	Cookies selection, Brown Betty teapot, 2 x 100g black tea	Brown Betty teapot, 2 x 100g black tea	2 x 100g black tea		Minus 1 x 100g black tea			
Items in order		Cookies selection	Cookies selection, Brown Betty teapot	Cookies selection, Brown Betty teapot, 2 x 100g tea	Cookies selection, Brown Betty teapot, 2 x 100g black tea	Cookies selection, Brown Betty teapot, 1 x 100g black tea	Cookies selection, Brown Betty teapot, 1 x 100g black tea	
Items subtotal	$0	$60	$76.50	$90	$90	$83.25	$83.25	$83.25
Shipping price	$0	?	?	?	$30.55	$30.55	$30.55	$30.55
Total price	$0	?	?	?	$120.55	$113.80	$113.80	$113.80
Items purchased								Cookies selection, Brown Betty teapot, 1 x 100g black tea

SDP1a-U's process structures support **abstract behaviours**, which are defined as "planning," "controlling," "perceiving," and "executing."

SDP1a-U's abstract structures are embodied by its **physical structures**, which support **physical behaviours**—that is, "clicking," "keying," and "searching."

Computer Model

SDP1a-C, like SDP1a-U, comprises both **representation structures** and **process structures**, which support **abstract behaviours**. The **abstract structures** are embodied by **physical structures**, for example, memory, processors, which support **physical behaviours**; see Figure 8.3.

The computer model abstract and physical structures and behaviours are not operationalised fully. The design problem allows for an increase in computer costs and as technological development is proceeding at a rapid pace, the issue of high computer costs is not considered critical.

128 8. CYCLE 1 DEVELOPMENT OF INITIAL HCI ENGINEERING DESIGN PRINCIPLES: B-TO-B

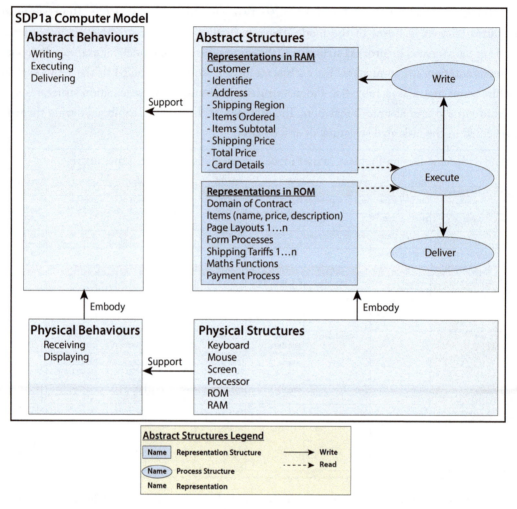

Figure 8.3: Specific Design Problem 1a computer model (following Cummaford, 2007).

Category Mapping between Models

The user model, computer model, and product goal contain concepts, which appear similar, but are not co-extensive. These concepts are summarised in the category mapping table; see Table 8.9.

Table 8.9: Specific Design Problem 1a category mapping

Product Goal	User	Computer
Transfer ownership rights	Exchange goods for currency	Transfer ownership rights
Item instance	Specific item	Instance of SKU inventory item

8.2 CYCLE 1 CLASS DESIGN PROBLEM/CLASS DESIGN SOLUTION SPECIFICATION 129

The user's notion of purchasing an item may be naïve. They may consider the transaction to be a simple exchange of the item for the purchase price. However, the legal status of the transaction may differ from the user's naïve view. The user may also have a naïve view of the item, conceptualising it as a specific instance of that item (for example, a particular teapot). However, the vendor conceptualises the item as being an instance of a specific product line item (known as a Stock Keeping Unit or SKU).

Task-Goal Structure

SDP1a-U and SDP1a-C interact to achieve the product goal, expressed as task goals. The user and computer behaviours, which interact to achieve the task goals, are specified in the **task-goal structure**, an excerpt of which appears in Table 8.10.

Table 8.10: Specific Design Problem 1a task goal structure t3 "add 2 units of black tea"

R	U/C	Task description	Dom trans.	UB(abstract)	PI	Co	En	Ex	UB(physical)	Search	click	key	CB (abstract)	CB (physical)
1	C	Display [S5]: category index page]							Display: link to sub category				Execute: form process. Deliver: S5	Display: S5 (category)
2	U	Read page, select category		Encode page. Control x2: get next item to purchase, choose category. Execute action		2	1	1	Search screen, click					
3	C	Display product listing page [S6]											Execute: form process. Deliver: S6	Receive: link. Display: S6 (product information, add to cart button)
4	U	Read page, find button, press 'add to cart'		Encode page. Control x2: get next item to purchase, choose item. Execute action		2	1	1	Search screen. Click on 'add to cart'	1	1			
5	C	Display shopping cart page [S3.3]											Write: items ordered, items subtotal. Execute: form process. Deliver: S3.3	Receive: add to cart command. Display: S3.3 (items in order, change quantity textbox, change quantity button)
6		Read page, click into quantity text box. Press delete, Type 2. Press Update Quantity button		Encode page. Control: select correct action x 2 (enter new quantity, and press update button). Execute action x 2.		2	1	2	Search screen. Click into text box, press delete, press 2, press 'update quantity'	1	2	2		
7		Display updated shopping cart page [S3.4]											Write: items ordered, items subtotal. Execute: form process. Deliver: S3.4	Receive: update quantity command. Display: S3.4 (items in order, goods subtotal)
8		Confirm item added to order		Encode page. Plan: update 'items to purchase' 'items in order' 'item subtotal'. Control: understand that item purchase task completed	3	1	1		Search screen.	1				
					3	7	4	4		3	3	2		

130   8. CYCLE 1 DEVELOPMENT OF INITIAL HCI ENGINEERING DESIGN PRINCIPLES: B-TO-B

Performance

Actual Performance

The task descriptions, contained in the task goal structure, are developed normatively, to ensure that the data reflect **actual user costs**, associated with error-free task completion. The user costs can be compared between the specific design problem and specific design solution. Neither is affected by user errors, which may impact the costs totals (for example, selecting an incorrect link and then returning would increase both abstract and physical user behaviours).

Actual task quality is measured empirically, as the percentage of users, who completed each task and the mean time taken. Both user costs and task quality are represented in the Specific Design Problem1a Costs Matrix; see Table 8.11.

Table 8.11: Specific Design Problem 1a costs matrix

		T1	T2	T3	T4	T5	T6	T7	Totals	Ordering
Abstract behaviours	Plan	3	3	3	2	3	2	5	21	
	Control	3	6	7	7	3	5	8	39	
	Encode	2	4	4	4	2	4	6	26	
	Execute	1	3	4	6	2	3	5	24	
	Total	**9**	**16**	**18**	**19**	**10**	**14**	**24**	**110**	
Physical behaviours	Search	2	4	3	4	2	4	5	24	
	Click	1	4	3	18	2	3	10	41	
	Keystroke	0	12	2	76	2	0	16	108	
	Total	**3**	**20**	**8**	**98**	**6**	**7**	**31**	**173**	
Task completion	U1	1	1	1	0	1	1	1	6	DS then DP
	U2	1	1	1	1	1	1	1	7	DP then DS
	U3	1	1	1	1	1	1	1	7	DP then DS
	U4	1	1	1	1	1	1	1	7	DS then DP
	U5	1	1	1	0	1	0	1	5	DP then DS
	U6	1	1	1	1	1	0	1	6	DS then DP
	Total	**6**	**6**	**6**	**4**	**6**	**4**	**6**	**38**	
Time to complete	U1	8	41	58	70	32	38	18	265	
	U2	6	114	64	107	20	37	44	392	
	U3	5	29	41	72	18	37	35	237	
	U4	9	61	53	163	12	26	51	375	
	U5	5	21	19	20	17	5	81	168	
	U6	4	22	34	112	19	36	57	284	
	Total	37	288	269	544	118	179	286	1721	
	Total / 6	**6.17**	**48**	**44.8**	**90.7**	**19.7**	**29.8**	**47.7**	**286.8**	

Desired Performance

The statement of desired performance shown here is taken from the Specific Design Problem 1a Product Goal.

Task Quality

All instances, which satisfy the pre-purchase requirements, should result in the product goal being achieved.

Interactive Worksystem Costs

User costs should be acceptable, and lower than completing the transaction via the specific instances of this class of transaction system, tested during Specific Design Problem 1a construction.

Any increase in computer costs is acceptable. The design solution must be implementable using existing technologies (that is, setup costs must not include development of new technologies).

Specific Design Problem 1b

The operationalisation of Specific Design Problem 1b is similar to that of Specific Design Problem 1a. A separate report, then, is not considered to be required.

8.2.3   STAGE 2: SPECIFY CLASS DESIGN PROBLEM

Following specification of Specific Design Problem 1a and Specific Design Problem 1b, commonalities are abstracted to construct the class design problem. This abstraction comprises common aspects of the specific design problems, to provide an initial class design problem expression. The class design problem, domain model, product goal, task goal structure, user model, and computer model are all constructed by abstraction. As it is not possible to test the class design problem empirically, as there are no class-level users as such, Pa for the class design problem is derived from the specific design problems tested.

Domain and Product Goal

Class Design Problem 1 is a class of transaction systems supporting the exchange of homogeneous physical goods for currency, which do not achieve a stated desired performance. The class task involves purchasing four items. During the task, one of the items had to be removed from the order, to satisfy the user's financial constraint. The task was: T1: Order 1 unit of Item 1 from Homepage; T2: Search for Item 2 and order 1 unit; T3: Order 2 units of Item 3; T4: Find the total price for the order (that is, total price, including shipping); T5: Delete 1 unit of Item 3 from the order; T6: Find the new total price for the order (that is, total price, including shipping); and T7: Complete the purchase (that is, pay for the order).

The **class domain model** (termed "CDP1-D") is shown in Figure 8.4 and comprises: the user; computer (e-shop); and the goods to be purchased. The domain model contains abstract objects, which are embodied in **physical objects**.

The **affordant domain attributes** (shown in bold in the model) are changed by the worksystem to achieve the product goal (see Table 8.12). The **dispositional domain attributes** must have the values specified in the product goal for the work to be effected (see Table 8.13). The class domain model does not contain specific values for the object attribute values. The requirements for the object attribute values are specified in the product goal. The **Product Goal** (termed "CDP1-PG") specifies the required values for the dispositional domain object attributes, and a specification of the affordant domain attribute value transformations that comprise the work.

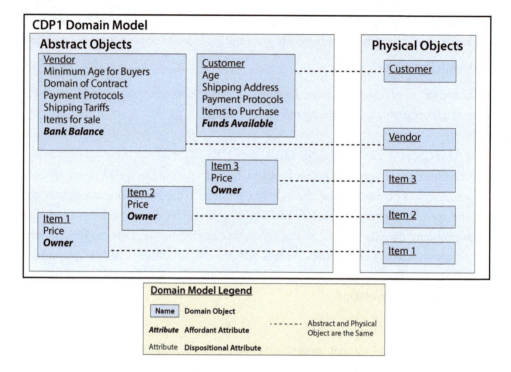

Figure 8.4: Class Design Problem 1 domain model (CDP1-D) (following Cummaford, 2007).

Table 8.12: CDP1-PG: dispositional object attribute value requirements	
Prepurchase Requirements	**Achieved?**
Customer: age [age] >= Vendor: minimum age for buyers [age]	Yes
Customer: shipping address [address] is a member of vendor: domain of contract [{set}]	Yes
Customer: Payment protocols [{set}] and Vendor: Payment protocols [{set}] must contain a common item	Yes
Customer: items to purchase [{set}] must be members of Vendor: items for sale [{set}]	Yes
Customer: funds available [amount] must be >= sum of price for every item in user: items to purchase, plus shipping costs as determined using vendor: shipping tariffs	Yes

8.2 CYCLE 1 CLASS DESIGN PROBLEM/CLASS DESIGN SOLUTION SPECIFICATION   133

Table 8.13: CDP1-PG: affordant object attribute value transformations

Domain Object: Attribute [value]	Start State	End State
Item1: owner [value]	Vendor	Customer
Item2: owner [value]	Vendor	Customer
Item3: owner [value]	Vendor	Customer
Customer: funds available [amount]	Amount	Amount minus sum of price for every item in user: items to purchase, plus shipping costs as determined using vendor: shipping tariffs
Vendor: bank balance[amount]	Amount	Amount plus sum of price for every item in user: items to purchase, plus shipping costs as determined using vendor: shipping tariffs

Class Worksystem

The class worksystem comprises a class user, termed "CDP1-U" and shown in Figure 8.4, who interacts with a class computer, termed "CDP1-C" and shown in Figure 8.5. CDP1-U and CDP1-C both comprise Representation Structures and Process Structures. Both CDP1-U and CDP1-C were abstracted from the commonalities between the respective specific user and computer models in Specific Design Problem 1a and Specific Design Problem 1b.

Class User Model (CDP1-U)

CDP1-U comprises two types of abstract structure. **Representation Structures** (shown in boxes in the model, see Figure 8.5) have particular states, for example, "items ordered," which are transformed by **process structures** (shown in ovals in the model, for example, "Encode"). The representation structure states for each stage of the work are detailed in the representation structure states matrix—see Table 8.14. Process structure activation of representation states is assumed to support user abstract behaviours. The user abstract behaviours, incurred during the work, are specified in the task goal structure, described later.

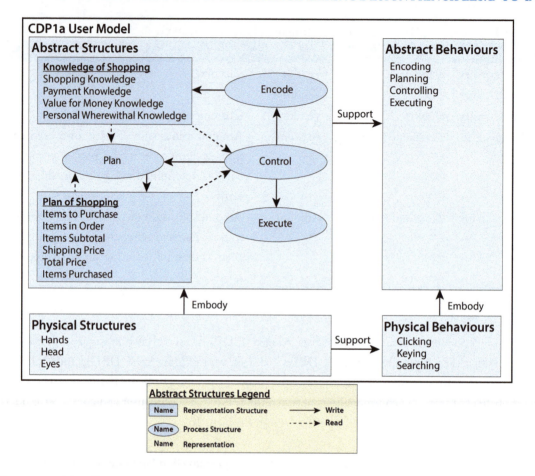

Figure 8.5: Class Design 1 User Model (CDP1-U) (following Cummaford, 2007).

The process structures in CDP1-U support **abstract behaviours**, which are defined as "planning," "controlling," "perceiving," and "executing."

The abstract structures in CDP1-U are embodied by its **physical structures**, which also support **physical behaviours**, that is, "clicking," "keying," and "searching."

8.2 CYCLE 1 CLASS DESIGN PROBLEM/CLASS DESIGN SOLUTION SPECIFICATION

Table 8.14: CDP1-U representation structure states matrix

	Start	After T1	After T2	After T3	After T4	After T5	After T6	After T7
Abstract Structures								
Shopping knowledge	Starting state	Plus T1 increment	Plus T2 increment	Plus T3 increment	Plus T4 increment	Plus T5 increment	Plus T6 increment	Plus T7 increment
Payment knowledge	Starting state							Plus T7 increment
Value for money knowledge	Starting state	Plus T1 increment	Plus T2 increment	Plus T3 increment	Plus T4 increment	Plus T5 increment	Plus T6 increment	Plus T7 increment
Personal wherewithal knowledge	Starting state							Plus T7 increment
Plan for shopping								
Items to purchase	P1, P2, 2xP3	P2, 2xP3	2xP3		Minus 1xP3			
Items in order		P1	P1, P2	P1, P2, 2xP3	P1, P2, 2xP3	P1, P2, 1xP3	P1, P2, 1xP3	
Items subtotal	£0	P1cost	P1cost + P2cost	P1cost + P2cost + 2xP3cost	P1cost + P2cost + 2xP3cost	P1cost + P2cost + P3cost	P1cost + P2cost + P3cost	P1cost + P2cost + P3cost
Shipping price	£0	?	?	?	Shipcost	?	Shipcost	Shipcost
Total price	£0	?	?	?	P1cost + P2cost + 2xP3cost + Shipcost	?	P1cost + P2cost + P3cost + Shipcost	P1cost + P2cost + P3cost + Shipcost
Items purchased								P1, P2, 1xP3

CDP1-C Model

CDP1-C, like CDP1-U, comprises both **representation structures** and **process structures**, which support **abstract behaviours**; see Figure 8.6. The **abstract structures** are embodied by **physical structures**, for example, memory, processors, which support **physical behaviours**.

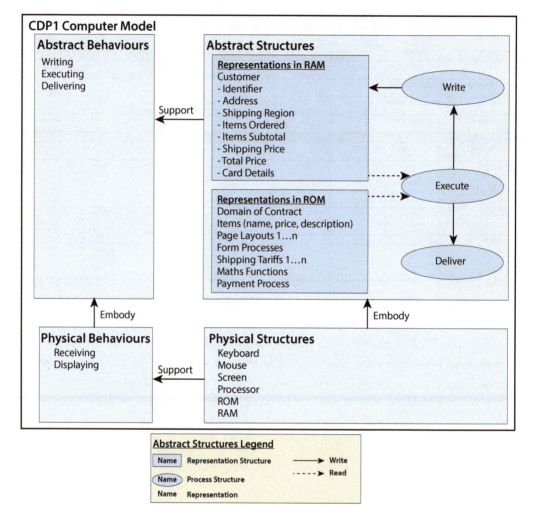

Figure 8.6: Class Design Problem 1 Computer Model (CDP1-C) (following Cummaford, 2007).

Category Mapping Between Models

CDP1-U and CDP1-C interact to achieve the domain transformations specified in CDP1-PG. However, their models of the work to be performed are not co-extensive with the domain transformations specified in CDP1-PG. The mappings between the concepts in CDP1-PPG and the worksystem models CDP1-U and CDP1-C are summarised in the category mapping table, shown in Table 8.15.

8.2 CYCLE 1 CLASS DESIGN PROBLEM/CLASS DESIGN SOLUTION SPECIFICATION 137

Table 8.15: Class Design Problem 1 category mapping

CDP1-PG	CDP1-U	CDP1-C
Transfer ownership rights	Buy	Transfer ownership rights
Item instance	Specific item	Instance of SKU inventory item

Task-Goal Structure

CDP1-U and CDP1-C interact to effect a number of task goals in order to achieve CDP-PG. The user and computer behaviours, which interact to achieve the task goals, are specified in the **class Task-Goal Structure** (termed "CDP1-TGS"). An excerpt from CDP1-TGS is presented in Table 8.16. The specific behaviours in SDP1a and SDP1b, at the device level of description, are not sufficiently similar to support generification of commonalities. The total number of each behaviour required to complete the tasks in the TGS-class is derived by taking the mean value for each behaviour across SDP1a and SDP1b. These values are represented in the CDP1 Costs Matrix, specified in Table 8.17.

Table 8.16: CDP1-TGS: T3 "Order 2 Units of Item 3"

R	U/C	Task description	Dom trans.	UB(abstract)	UB(physical)	CB (abstract)	CB (physical)
1	C	Display category index page [S5]				Execute: form process. Deliver: S5	Display: S5 (category)
2	U	Read page, select category		Encode: page. Control x2: get next item to purchase, choose category. Execute: action	Search screen, click		
3	C	Display product listing page [S6]				Execute: form process. Deliver: S6	Receive: link. Display: S6 (product information, add to cart button)
4	U	Read page, find button, press 'add to cart'		Encode: page. Control x2: get next item to purchase, choose item. Execute: action	Search screen. Click on 'add to cart'		
5	C	Display shopping cart page [S3.3]				Write: items ordered, items subtotal. Execute: form process. Deliver: S3.3	Receive: add to cart command. Display: S3.3 (items in order, change quantity textbox, change quantity button)
6	U	Read page, click into quantity text box. Press delete, Type 2. Press Update Quantity button		Encode: page. Control x2: select correct action x 2 (enter new quantity, and press update button). Execute x2: enter quantity, press update button	Search screen. Click into text box, press delete, press 2, click 'update quantity'		
7	C	Display updated shopping cart page [S3.4]				Write: items ordered, items subtotal. Execute: form process. Deliver: S3.4	Receive: update quantity command. Display: S3.4 (items in order, goods subtotal)
8	U	Confirm item added to order		Encode: page. Plan x3: update 'items to purchase' 'items in order' 'item subtotal'. Control: understand that item purchase task completed	Search screen		

Performance

Actual Performance

The task descriptions contained in CDP1-TGS are abstracted from the commonalities between the task goal structures in SDP1a and SDP1b. CDP1-TGS is therefore computer independent, as the computer-specific aspects of the SDP TGSs are not common between SDP1a and SDP1b. The **class actual performance** is derived from the mean number of each behaviour contained in the task goal structures for SDP1a and SDP1b. These mean values are represented in the CDP1 Costs Matrix (termed "CM1-class").

The **class actual task quality** is derived from the actual task quality of SDP1a and SDP1b.

Both user costs and task quality are expressed in the CDP1 Costs Matrix (termed "CDP1-CM") shown in Table 8.17.

Table 8.17: CDP1 Costs Matrix (CDP1-CM)

		T1	T2	T3	T4	T5	T6	T7	Totals
Abstract behaviours	Plan	3	3	3	2	3	2	4	20
	Control	3.5	6	7	7	3	5.5	8.5	40.5
	Encode	2.5	4	4	5.5	2	5.5	5	28.5
	Execute	1.5	3	4	4.5	2	3	6.5	24.5
	Total	**10.5**	**16**	**18**	**19**	**10**	**16**	**24**	**113.5**
Physical behaviours	Search	2.5	4	3	5.5	2	5.5	4.5	27
	Click	1.5	6	3	10.5	2	3	17.5	43.5
	Keystroke	0	10	2	38	2	0	54	106
	Total	**4**	**20**	**8**	**54**	**6**	**8.5**	**76**	**176.5**
Task completion	**Percentage**	**100.0%**	**100.0%**	**100.0%**	**33.3%**	**100.0%**	**33.3%**	**100.0%**	**81.0%**
Time to complete (secs)	**Average**	**18.917**	**36.75**	**43.417**	**114.17**	**17.833**	**32.5**	**58.5**	**322.08**

Desired Performance

Task Quality

All instances which satisfy the pre-purchase requirements should result in the product goal being achieved.

IWS costs

User costs should be acceptable and lower than the class actual costs, derived during specification of CDP1.

Any increase in computer costs is acceptable. The design solution must be implementable using existing technologies (that is, setup costs must not include development of new technologies).

8.2.4 STAGE 3: EVALUATE CLASS DESIGN PROBLEM

The e-shops evaluated during the specification of SDP1a and SDP1b exhibit different behaviours at the device level of description. For example, shipping prices required the user to find international shipping rates and calculate the likely shipping cost in SDP1a. While the user had to register to find shipping costs in SDP1b. These differences result in a similar product goal being achieved, but with different actual performances. However, aspects of the work resulting in high workload are similar. They are therefore included in the class design problem. The latter's user model (CDP1-U) and domain model (CDP1-D) are operationalised analytically, to check that the task goal structure in each specific design problem could be achieved. CDP1-U contains appropriate behaviours to achieve the task goal structure in each of the specific design problems tested. The class design problem is retained.

8.2.5 STAGE 4: SPECIFY CLASS DESIGN SOLUTION

Best-practice design knowledge and guidelines were recruited to develop CDS1. In addition, behaviours resulting in low task quality or high user costs were identified and re-engineered to increase potentially achievable performance. The development is summarised earlier. Here, the CDS1 components are presented.

Domain and Product Goal

CDS1 is a class of transaction systems supporting the exchange of homogeneous physical goods for currency, which achieve a stated desired performance. The class task involves purchasing four items. During the task, one of the items has to be removed from the order, to satisfy the user's financial constraint. The task scenario comprises: T1: Order 1 unit of Item 1 from Homepage; T2: Search for Item 2 and order 1 unit; T3: Order 2 units of Item 3; T4: Find the total price for the order (that is, total price, including shipping); T5: Delete 1 unit of Item 3 from the order; T6: Find the new total price for the order (that is, total price, including shipping); and T7: Complete the purchase (that is, pay for the order).

The **class domain model** (termed CDS1-D) comprises the user, computer (e-shop), and the goods to be purchased. The domain model contains **abstract objects**, which are embodied in **physical objects** (see Figure 8.7).

The **affordant domain attributes** (shown in bold in the model) are changed by the worksystem to achieve the Product Goal. The **dispositional domain attributes** must have the values specified in the Product Goal, in order for it to be possible for the work to be effected. The class domain model does not contain specific values for the object attribute values. The requirements for the object attribute values are specified in the Product Goal. The **Product Goal** (termed CDS1-PG) specifies the required values for the dispositional domain object attributes (Table 8.18) and a specification of the affordant domain attribute value transformations that comprise the work (Table 8.19).

140 8. CYCLE 1 DEVELOPMENT OF INITIAL HCI ENGINEERING DESIGN PRINCIPLES: B-TO-B

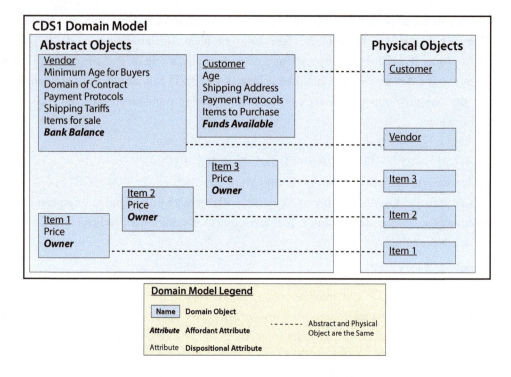

Figure 8.7: CDS1 Domain Model (CDS1-D) (following Cummaford, 2007).

Table 8.18: CDS1-PG: dispositional object attribute value requirements	
Prepurchase Requirements	**Achieved?**
Customer: age [age] >= Vendor: minimum age for buyers [age]	Yes
Customer: shipping address [address] is a member of vendor: domain of contract [{set}]	Yes
Customer: Payment protocols [{set}] and Vendor: Payment protocols [{set}] must contain a common item	Yes
Customer: items to purchase [{set}] must be members of Vendor: items for sale [{set}]	Yes
Customer: funds available [amount] must be >= sum of price for every item in user: items to purchase, plus shipping costs as determined using vendor: shipping tariffs	Yes

8.2 CYCLE 1 CLASS DESIGN PROBLEM/CLASS DESIGN SOLUTION SPECIFICATION   141

Table 8.19: CDS1-PG: affordant object attribute value transformations

Domain Object: Attribute [value]	Start State	End State
Item1: owner [value]	Vendor	Customer
Item2: owner [value]	Vendor	Customer
Item3: owner [value]	Vendor	Customer
Customer: funds available [amount]	Amount	Amount minus sum of price for every item in user: items to purchase, plus shipping costs as determined using vendor: shipping tariffs
Vendor: bank balance[amount]	Amount	Amount plus sum of price for every item in user: items to purchase, plus shipping costs as determined using vendor: shipping tariffs

Class Worksystem

The class worksystem comprises a class user, termed "CDS1-U" (Figure 8.8), who interacts with a class computer, termed "CDS1-C." CDS1-U and CDS1-C both comprise Representation Structures and Process Structures. Both CDS1-U and CDS1-C are abstracted from the commonalities between the respective specific user and computer models in SDS1a and SDS1b.

Class user model (CDS1-U)

CDS1-U comprises two types of abstract structure. **Representation structures** (shown in boxes in the model) have particular states, for example, "items ordered," which are transformed by **process structures** (shown in ovals in the model). The representation structure states for each stage of the work are detailed in the representation structure states matrix (Table 8.20). Process structure activation of representation states is assumed to constitute user abstract behaviours. The user abstract behaviours exhibited during the work are specified in the task goal structure and discussed later.

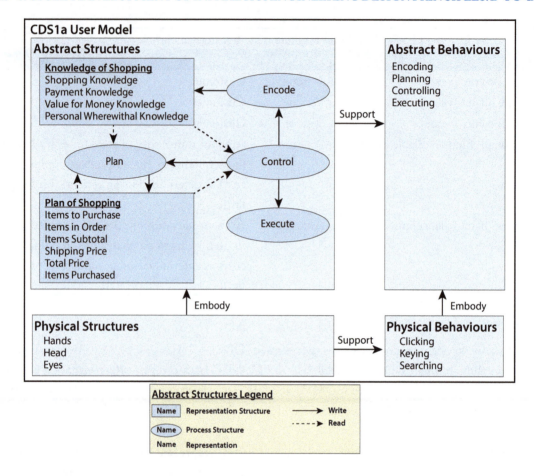

Figure 8.8: CDS1 User Model (CDS1-U) (following Cummaford, 2007).

The process structures in CDS1-U support **abstract behaviours**, which are defined as planning, controlling, perceiving, and executing.

The abstract structures in CDS1-U are embodied by its **physical structures**, which also support **physical behaviours**, that is, clicking, keying, and searching.

8.2 CYCLE 1 CLASS DESIGN PROBLEM/CLASS DESIGN SOLUTION SPECIFICATION   143

Table 8.20: CDS1-U representation structure states matrix

	Start	After T1	After T2	After T3	After T4	After T5	After T6	After T7
Abstract Structures								
Shopping knowledge	Starting state	Plus T1 increment	Plus T2 increment	Plus T3 increment	Plus T4 increment	Plus T5 increment	Plus T6 increment	Plus T7 increment
Payment knowledge	Starting state							Plus T7 increment
Value for money knowledge	Starting state	Plus T1 increment	Plus T2 increment	Plus T3 increment	Plus T4 increment	Plus T5 increment	Plus T6 increment	Plus T7 increment
Personal wherewithal knowledge	Starting state							Plus T7 increment
Plan for shopping								
Items to purchase	P1, P2, 2xP3	P2, 2xP3	2xP3		Minus 1xP3			
Items in order		P1	P1, P2	P1, P2, 2xP3	P1, P2, 2xP3	P1, P2, 1xP3	P1, P2, 1xP3	
Items subtotal	£0	P1cost	P1cost + P2cost	P1cost + P2cost + 2xP3cost	P1cost + P2cost + 2xP3cost	P1cost + P2cost + P3cost	P1cost + P2cost + P3cost	P1cost + P2cost + P3cost
Shipping price	£0	?	?	?	Shipcost	Shipcost	Shipcost	Shipcost
Total price	£0	?	?	?	P1cost + P2cost + 2xP3cost + Shipcost	P1cost + P2cost + P3cost + Shipcost	P1cost + P2cost + P3cost + Shipcost	P1cost + P2cost + P3cost + Shipcost
Items purchased								P1, P2, 1xP3

CDS1-C

CDS1-C (see Figure 8.9), like CDS1-U, comprises both **representation structures** and **process structures**, which support **abstract behaviours**. The **abstract structures** are embodied by **physical structures**, for example, memory, processors, which support **physical behaviours**.

Category mapping between models

CDS1-U and CDS1-C interact to achieve the domain transformations specified in CDS1-PG. However, their model of the work to be performed is not co-extensive with the domain transformations specified in CDS1-PG. The mappings between the concepts in CDS1-PG and the worksystem models CDS1-U and CDS1-C are summarised in the category mapping table shown in Table 8.21.

Table 8.21: CDS1 category mapping table

CDS1-PG	CDS1-U	CDS1-C
Transfer ownership rights	Buy	Transfer ownership rights
Item instance	Specific item	Instance of SKU inventory item

Task-Goal Structure

CDS1-U and CDS1-C interact to achieve CDS1-PG. To do so, they achieve task goals. The user and computer behaviours are specified in the **class Task-Goal Structure** (termed CDS1-TGS). An excerpt from CDS1-TGS is presented in Table 8.22. TGS-class specifies the CDS1-U and CDS1-C task descriptions, but does not contain behaviours. The specific behaviours in SDS1a and SDS1b are not sufficiently similar to abstract commonalities. The total number of each

144   8. CYCLE 1 DEVELOPMENT OF INITIAL HCI ENGINEERING DESIGN PRINCIPLES: B-TO-B

behaviour required to complete the tasks in TGS-class is derived by taking the mean value for each behaviour across SDS1a and SDS1b. These values are represented in the class Costs Matrix, shown in Table 8.23.

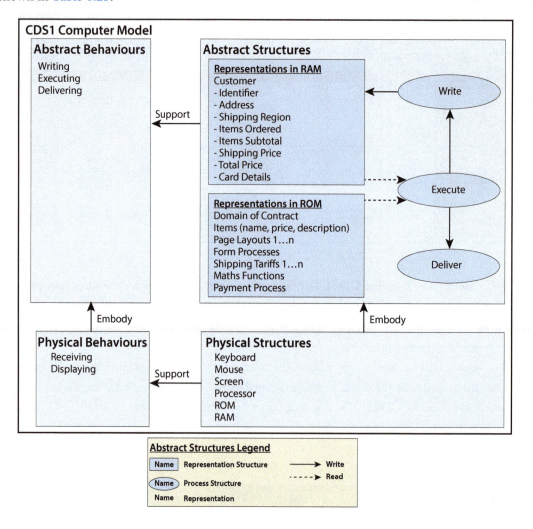

Figure 8.9: CDS1 Computer Model (CDS1-C) (following Cummaford, 2007).

8.2 CYCLE 1 CLASS DESIGN PROBLEM/CLASS DESIGN SOLUTION SPECIFICATION   145

Table 8.22: CDS1-TGS T3 "Order 2 Units of Item 3"

R	U/C	Task description	Dom trans.	UB(abstract)	UB(physical)	CB(abstract)	CB (physical)
1	C	Display category index page [S6]			Display: link to sub category	Execute: form process. Deliver: S6	Display: S6 (category)
2	U	Read page, select category		Encode: page. Control x2: get next item to purchase, choose category. Execute: action	Search screen, click		
3	C	Display product listing page [S7]				Execute: form process. Deliver: S7	Receive: link. Display: S7 (product information, add to basket button)
4	U	Read page, find button, press 'add to basket' twice		Encode: page. Control x2: get next item to purchase, choose item. Execute: action	Search screen. Click x 2		
5	C	Display page with updated shopping basket [S9]				Write: items ordered, items subtotal, shipping price, total price. Execute: form process. Deliver: S9	Receive: add to basket command x 2. Display: S9 (last item added, item subtotal,ship cost, total price)
6	U	Confirm item added to order		Encode: page. Plan x3: update 'items to purchase' 'items in order' 'item subtotal'. Control: understand that item purchase task completed	Search screen		

8.2.6   STAGE 6: EVALUATE CLASS DESIGN SOLUTION

Actual Performance

The task descriptions contained in CDS1-TGS are abstracted from the commonalities between the TGSs in SDS1a and SDS1b. CDS1-TGS is therefore computer independent, as the computer-specific aspects of the SDS TGSs are not common between SDS1a and SDS1b. The **class actual performance** is derived from the mean number of each behaviour contained in the TGSs for SDS1a and SDS1b. These mean values are represented in the CDS1 Costs Matrix (termed CDS1-CM) (see Table 8.23).

The **class actual task quality** is derived from the actual task quality of SDS1a and SDS1b. Both user costs and task quality are expressed in CDS1-CM, as shown in Table 8.23.

Table 8.23: CDS1 costs matrix (CDS1-CM)

CDS1 (SDS1a + SDS1b)/2		T1	T2	T3	T4	T5	T6	T7	Totals
Abstract behaviours	Plan	5	3	3	2	4	2	3	22
	Control	4	5	5	0	3	0	12	29
	Encode	4	2	3	1	3	1	5	19
	Execute	2	2	2	0	2	0	10	18
	Total	**15**	**12**	**13**	**3**	**12**	**3**	**30**	**88**
Physical behaviours	Search	3	2	3	0	3	0	5	16
	Click	3	3	3	0	2	0	27	38
	Keystroke	0	10	0	0	0	0	107	115
	Total	**6**	**13**	**6**	**0**	**5**	**0**	**139**	**169**
Task completion	Percentage	100.0%	100.0%	100.0%	100.0%	100.0%	100.0%	100.0%	100.0%
Time to complete (secs)	Average	10.5	16.333	20.083	19.583	13.583	5.0833	57.417	142.583333

Comparison of Pa and Pd

Task Quality

All instances, which satisfy the pre-purchase requirements, should result in the product goal being achieved. Actual performance is 100% task completion, therefore for CDS1 Pa = Pd, and can be considered a design solution.

IWS Costs

User costs are acceptable and lower than the class actual costs, derived during specification of CDP1. A comparison of the cost differences between CDP1 and CDS1 appears in the comparison Costs Matrix shown in Table 8.24.

Negative values indicate a reduction in costs between the CDP and CDS. Where costs increased between CDP and CDS, they are shown in red. Task completion rates show the increase in task completion percentages, rather than reductions in task failures to maintain consistency with the labelling of the Costs Matrices.

Table 8.24: Comparison of CDP1/CDS1 costs matrices

		T1	T2	T3	T4	T5	T6	T7	totals
Abstract behaviours	Plan	2	0	0	0	1	0	-1	2
	Control	0.5	-1	-2	-7	0	-5.5	3.5	-11.5
	Encode	1.5	-2	-1	-4.5	1	-4.5	0	-9.5
	Execute	0.5	-1	-2	-4.5	0	-3	3.5	-6.5
	Total	4.5	-4	-5	-16	2	-13	6	-25.5
Physical behaviours	Search	0.5	-2	0	-5.5	1	-5.5	0.5	-11
	Click	1.5	-3	0	-10.5	0	-3	9.5	-5.5
	Keystroke	0	0	-2	-38	-2	0	53	11
	Total	2	-7	-2	-54	-1	-8.5	63	-7.5
Task completion	Increase	0.00%	0.00%	0.00%	66.67%	0.00%	66.67%	0.00%	19.05%
Time to complete	Average	-8.4	-20.4	-23.3	-94.6	-4.3	-27.4	-1.1	-179.5

The design solution is implementable using existing technologies, and so can be considered a valid design solution.

The user ratings of workload and comparative workload support the conclusion that less workload is required to complete the tasks with the specific design solution systems. The latter, and so by extension the class design solution, are therefore considered acceptable. They fulfil the requirement of the product goals for the class design solution and specific design solutions systems to result in lower user costs for task completion.

The CDS is thus considered acceptable and retained.

REVIEW

The chapter reports the Cycle 1 class design problem and class design solution for physical goods transaction systems. The cycle follows the development process of, and the method for, identifying, class design problems and their class design solutions specified earlier.

8.3 PRACTICE ASSIGNMENT

8.3.1 GENERAL

Read § 8.1, concerning Cycle 1 development.

- Check Cycle 1 informally for completeness and coherence, as required by the case study of business-to-consumer electronic commerce.

- The aim of the work assignment is for you to become sufficiently familiar with Cycle 1 to apply it subsequently and as appropriate to a different domain of application, as in Practice Scenarios 8.1–2.

Hints and Tips

Difficult to get started?

Re-read the assignment task carefully.

- Make written notes and in particular list the sections, while re-reading § 8.1.

- Think about how the sections might be applied to describe a novel domain of application.

- Re-attempt the assignment.

Test

List from memory as many sections of Cycle 1 as you can.

Read § 8.2, concerning Cycle 1 development.

- Complete as for the previous section starting Read § 8.1.

8.3.2 PRACTICE SCENARIOS

Practice Scenario 8.1: Applying Cycle 1 Development to an Additional Domain of Application

Select an additional domain of application. The domain should be other than that of business-to-consumer electronic commerce.

- Apply the Cycle 1 development for business-to-consumer electronic commerce (see § 8.1) to the novel domain of application. The description can only be of the most general kind - that is at the level of the section. However, even consideration at this high level orient the researcher towards application of the Cycle 1 development to novel domains of application. The latter are as might be required subsequently by their own work. The practice scenario is intended to help bridge this gap.

Practice Scenario 8.2: Applying Cycle 1 Class Design Problem/Class Design Solution Specification to an Additional Domain of Application

Select the same additional domain of application, as used in Practice Scenario 8.1. The domain should be other than that of business-to-consumer electronic commerce.

- Complete as for Practice Scenario 8.1.

CHAPTER 9

Cycle 2 Development of Initial HCI Engineering Design Principles for Business-to-Consumer Electronic Commerce

SUMMARY

This chapter reports Cycle 2 development of initial HCI-EDPs for the domain of business-to-consumer electronic commerce. The section comprises: Cycle 2 development and class design problem/ class design solution specification.

Space limitations preclude for Cycle 2 the same in-depth and detailed report made for Cycle 1. Thus, only an outline of Cycle 2 is presented here. The outline, however, is sufficient for readers' needs. A complete Cycle 2 report is available online. Given this, see Note [1] for different ways for readers to approach the chapter.

9.1 CYCLE 2 DEVELOPMENT

This section does for Cycle 2 development what 8.1 does for Cycle 1 and with the same structure. The content, however, is that of Cycle 2. The latter comprises the development of a class design problem and class design solution for electronic goods (information and software). The latter concern SMS-based sports news alerts (ManUtd.com), and mobile phone ringtones and software (Jamster.com), respectively.

9.1.1 INTRODUCTION

As in § 8.1.1, a class design solution, corresponding to the class design problem, is specified and instantiated for Cycle 2 as specific design solutions, corresponding to the two specific design problems selected.

9.1.2 SELECTION OF SYSTEMS FOR SPECIFIC DESIGN PROBLEM AND SPECIFIC DESIGN SOLUTION DEVELOPMENT

As in § 8.1.2, two electronic shops are selected for Cycle 2. They are: Specific Design Problem 2a—Manchester United Text Alerts Service (www.manutd.com) and Specific Design Problem 2b—Jamster (www.jamster.com) mobile phone ringtones, games, and screensavers for sale.

9.1.3 TESTING PROCEDURE

As in § 8.1.3, empirical testing is carried out for Cycle 2, comprising Set Up, Participants, Procedure, Testing Tasks, Calculation of User Costs; also a table of Criteria for Diagnosing User Abstract Behaviour, as in Table 8.1, but as required for Cycle 2.

9.1.4 SPECIFY SPECIFIC DESIGN PROBLEMS

As in § 8.1.4, the results from the Cycle 2 testing of Specific Design Problems 2a/2b, including: Specific Design Problem 2a; table for specific Design Problem 2a; Costs Matrix; Specific Design Problem 2b; and also a table for Specific Design Problem 2b Costs Matrix, as in Table 8.3, but all as required for Cycle 2.

9.1.5 SPECIFY CLASS DESIGN PROBLEM

As in § 8.1.5, including Cycle 2 Class Design Problem, comprising: Domain Model; Product Goal; Task-Goal Structure, and Interactive Worksystem Models (User Model and Computer Model).

9.1.6 EVALUATE CLASS DESIGN PROBLEM

As in § 8.1.6, evaluate Cycle 2 class design problem.

9.1.7 SPECIFY CLASS DESIGN SOLUTION

As in § 8.1.7, specify Cycle 2 class design solution.

9.1.8 SPECIFY SPECIFIC DESIGN SOLUTIONS

As in § 8.1.8, including Cycle 2 testing results.

9.1.9 EVALUATE CLASS DESIGN SOLUTION

As in § 8.1.9, including for Cycle 2: Class Design Problem 2 Costs Matrix, and Class Design Solution 2 Costs Matrix, as in Tables 8.4 and 8.5, but for Cycle 2.

9.2 CYCLE 2 CLASS DESIGN PROBLEM/CLASS DESIGN

SOLUTION SPECIFICATION

As in § 8.2, Cycle 2 Design Problem/Class Design Solution Specification.

9.2.1 INTRODUCTION

As in § 8.2.1, Cycle 2 introduction.

9.2.2 STAGE 1: SPECIFY SPECIFIC DESIGN PROBLEMS

As in § 8.2.2, including Cycle 2: Specific Design Problem 2a; Specific Design Problem 2a Domain Model; Specific Design Problem 2a Product Goal: Dispositional Object Attribute Value Requirements; Specific Design Problem 2a Product Goal: Affordant Object Attribute Value Transformations; Specific Design Problem 2a User Model; Specific Design Problem 2a User Model Representation Structure States Matrix; Specific Design Problem 2a Computer Model; Specific Design Problem 2a Category Mapping; Specific Design Problem 2a Task Goal Structure T3, and Specific Design Problem 2a Costs Matrix. These are all as in Figures 8.1–8.3 and Tables 8.1–8.11, but as required for Cycle 2.

The operationalisation of Specific Design Problem 2b is similar to that of Specific Design Problem 2a. A separate listing, then, is not considered to be required.

9.2.3 STAGE 2: SPECIFY CLASS DESIGN PROBLEM

As in § 8.2.3, including Cycle 2: Domain and Product Goal; Class Design Problem 2 Domain Model (CDP2-D) CDP2-PG: Dispositional Object Attribute Value Requiremets, CDP2-PG: Affordant Object Attribute Value Transformations, including: Class Worksystem, Class Design 2; User Model (CDP2-U); CDP2-U Representation Structure States Matrix, Class Design Problem 2; Computer Model (CDP2-C), Category Mapping Between Models, Class Design Problem 2 Category Mapping, Task-Goal Structure, CDP2-TGS: T3 "Order 2 units of item 3," Performance, CDP2 Costs Matrix (CDP2-CM). These are all as in Figures 8.4–8.6 and Tables 8.12–8.17, but as required for Cycle 2.

9.2.4 STAGE 3: EVALUATE CLASS DESIGN PROBLEM

As in § 8.2.4, Cycle 2 evaluate class design problem.

9.2.5 STAGE 4: SPECIFY CLASS DESIGN SOLUTION

As in § 8.2.4, including: Cycle 2: Domain and Product Goal, CDS2; Domain Model (CDS2-D), CDS2-PG: Dispositional Object Attribute Value Requirements, CDS2-PG: Affordant Object Attribute Value Transformations, including: Class Worksystem, CDS2: User Model (CDS2-U), CDS2-U Representation Structure States Matrix, CDS2; Computer Model (CDS2-C), CDS2

152 9. CYCLE 2 DEVELOPMENT OF INITIAL HCI ENGINEERING DESIGN PRINCIPLES: B-TO-B

Category Mapping Table, CDS2-TGS T3 "Order 2 units of item 3." These are all as in Figures 8.7–8.9 and Tables 8.18–8.22, but as required for Cycle 2.

9.2.6 STAGE 6: EVALUATE CLASS DESIGN SOLUTION

As in § 8.2.6, including Cycle 2: Actual Performance, CDS2 Costs Matrix (CDS2-CM), and Comparison of CDP2/CDS2 Costs Matrices, as in Tables 8.23 and 8.24 but required for Cycle 2.

Review

The chapter reports the Cycle 2 development of initial HCI-EDPs for the domain of business-to-consumer electronic commerce. The section comprises: Cycle 2 development, and class design problem/class design solution specification. Only an outline of Cycle 2 is presented, due to space limitations. As a result, see Note [1] for ways for readers to approach the chapter.

9.3 PRACTICE ASSIGNMENT

9.3.1 GENERAL

Read § 9.1, concerning Cycle 2 development.

- Using the main section titles § 9.1.1–§ 9.2.6 check the Cycle 2 development informally for completeness and coherence against Cycle 1, as required by the case study of business-to-consumer electronic commerce.

- The aim of the practice assignment is for you to become sufficiently familiar with Cycle 2 development to apply it subsequently and as appropriate to the different domain of application, selected for Practice Scenarios 9.1–2.

Hints and Tips

Difficult to get started?

Re-read the assignment task carefully.

- Make written notes and in particular list the main section headings, after re-reading § 9.1.

- Think about how the sections might be applied to describe Cycle 2 development of a novel domain of application.

- Re-attempt the assignment.

Test

List from memory as many of the section headings as you can.

Read § 9.2, concerning Cycle 2 class design problem/class design solution specification.

- Complete as for the previous section starting Read § 9.1.

9.3.2 PRACTICE SCENARIOS

Practice Scenario 9.1: Applying Cycle 2 Development to an Additional Domain of Application

Select an additional domain of application, other than that of business-to-consumer electronic commerce.

- Apply Cycle 2 development for business-to-consumer electronic commerce (see § 9.1 and main titles) to the novel domain of application. The description can only be of the most general kind—that is at the level of the section titles. However, even consideration at this high level can orient the researcher towards application of Cycle 2 development to novel domains of application. The latter are as might be required subsequently by their own work. The research design scenario is intended to help bridge this gap.

Practice Scenario 9.2: Applying Cycle 2 Class Design Problem/Class Design Solution Specification to an Additional Domain of Application

Select the same additional domain of application, as used in Practice Scenario 9.1. The domain should be other than that of business-to-consumer electronic commerce.

- Complete as for Practice Scenario 9.1.

Notes

[1] As stated in the Summary, space limitations preclude the same in-depth and detailed report for Cycle 2, as is presented in Chapter 8 for Cycle 1. A full Cycle 2 development report, however, is available elsewhere both on the publisher's website at https://bit.ly/Cycle2Report and on that of the authors at www.hciengineering.net/HCI-EDPs. Readers, thus have the following options:

Option 1. Read the full website version of Cycle 2 development after reading Chapter 8 and before reading Chapter 10. This option might best suit those reading the book primarily to understand its contents and whose grasp of Cycle 1 needs reinforcing.

Option 2. Read or just scan the outline version of Cycle 2 in Chapter 9. This option might best suit those readers who feel confident in their understanding of the Cycle 1 report and do not need to read a full Cycle 2 report. Section titles and figure/table prompts are sufficient for their needs.

Option 3. Follow the outline version of Cycle 2 and use it to illustrate a full report. The illustration is for the purpose of understanding. Chapters 8 and 9 have identical structures. The Cycle 2 outline includes all the Cycle 1 section titles and the associated figures and tables. Readers are, thus, supported in illustrating a full version of 2, based on Cycle 1. This option might best suit researchers either trying to replicate Cycle 2 or trying to apply the development to a new domain. Their illustration efforts can always be checked against the full version of Chapter 9, which is presented on the websites, cited earlier. This latter option might be considered a rather extensive "Practice Assignment." Any data, of course, could only be hypothetical. However, inventing such data would support understanding of the exercise.

Readers, of course, can mix and match options to suit their individual needs.

CHAPTER 10

Initial HCI Engineering Design Principles for Business-to-Consumer Electronic Commerce

SUMMARY

This chapter presents Cycle 1 and Cycle 2 acquisition of initial HCI-EDPs for business-to-consumer electronic commerce. The acquisition follows application of the associated specification method. The chapter comprises: the scope of the HCI-EDP; its specification; and its achievable performance. The identification of initial HCI-EDPs is reported.

10.1 HCI ENGINEERING DESIGN PRINCIPLE SPECIFICATION REQUIREMENTS

The HCI-EDP specification requirements comprise HCI-EDP: specification method; components; scope; specification; and achievable performance.

10.1.1 HCI ENGINEERING DESIGN PRINCIPLE SPECIFICATION METHOD

The method for specifying HCI-EDPs requires identification of commonalities between a class design problem and its class design solution. The latter constitute the scope of the HCI-EDP. Those aspects of the class design solution not included in the class design problem-class design solution commonalities, and the negation of those aspects of the class design problem not included in the class design problem-class design solution commonalities, are then used to define the prescriptive component of the principle. The latter's achievable performance is defined as the actual performance of the class design solution.

10.1.2 HCI ENGINEERING DESIGN PRINCIPLE COMPONENTS

An HCI-EDP comprises three components: a scope (which supports diagnosis); a specification (which supports prescription); and a class of achievable performance (which supports validation, leading to guarantee). The HCI-EDP embodies class design knowledge, derived from class design solutions. The method for principle construction, specified earlier, is used to organise the reporting of its acquisition.

10.1.3 HCI ENGINEERING DESIGN PRINCIPLE SCOPE

The HCI-EDP'scope defines the boundary of its applicability. The scope comprises a class of users and a class of computers. The latter interact to achieve a class of domain transformations within a class of domains. The engineering design problem scope is defined by generification of the commonalities between the class design problem and the class design solution. This ensures that sufficient components of the class design problem enable those of the class design solution to be operationalised.

An HCI-EDP is prescriptive class-level design knowledge, with a specified scope of application. The latter's scope enables comparison with that of other related HCI-EDPs, to determine their relative generality.

10.1.4 HCI ENGINEERING DESIGN PRINCIPLE SPECIFICATION

The HCI-EDP prescriptive design knowledge is synthesised from the non-common aspects of the class design solution and the class design problem. The relevant components of class design solution and class design problem are first compared to identify their non-commonalities. The latter are then used to construct the prescriptive component of the HCI-EDP. User and computer representation structure states are compared first, then user and computer behaviours. The computer physical structures, sufficient to support these user and computer behaviours, are referenced to the user and computer behaviours.

10.1.5 HCI ENGINEERING DESIGN PRINCIPLE ACHIEVABLE PERFORMANCE

The remaining components of the class design solution comprise the aspects, which, if operationalised for the HCI-EDP's scope components, achieve the level of performance, stated in the class design solution. The class design solution-only components are the foundation of the HCI-EDP. They provide a prescriptive specification of: a task goal structure; user and computer representation structure states; supported user and computer behaviours (assumed to be commensurate with process structure activations); and achievable performance, as task quality and worksystem costs.

10.2 HCI ENGINEERING DESIGN PRINCIPLES ACQUIRED IN CYCLE 1 DEVELOPMENT

The section comprises HCI-EDP: scope; specification, and achievable performance.

10.2.1 HCI ENGINEERING DESIGN PRINCIPLE SCOPE

The respective product goals, domain models, and user models are identical for both the class design problem and the class design solution. The models are thus recruited to the scope of the HCI-EDP.

The CDP1 (Class Design Problem1) and CDS1 (Class Design Solution1) user representation states have some commonalities. These are shown in Table 10.1. The remaining user representation states are not common, and so are not recruited to EDP1 scope.

Table 10.1: EDP1-U representation structure states matrix								
	Start	After T1	After T2	After T3	After T4	After T5	After T6	After T7
Abstract Structure								
Shopping knowledge	Starting State	Plus T1 Increment	Plus T2 Increment	Plus T3 Increment	Plus T4 Increment	Plus T5Increment	Plus T6 Increment	Plus T7 Increment
Payment knowledge	Starting State							Plus T7 Increment
Value for money knowledge	Starting State	Plus T1 Increment	Plus T2 Increment	Plus T3 Increment	Plus T4 Increment	Plus T5Increment	Plus T6 Increment	Plus T7 Increment
Personal wherewithal knowledge	Starting State							Plus T7 Increment
Plan for shopping								
Item to purchase	P1, P2, 2xP3	P2, 2xP3	2xP3		Minus 1xP3			
Items in order		P1	P1, P2	P1, P2, 2xP3	P1, P2, 2xP3	P1, P2, 1xP3	P1, P2, 1xP3	
Items subtotal	£0	P1 cost	P1cost + P2cost	P1cost + P2cost + 2xP3cost	P1cost + P2cost + 2xP3cost	P1cost + P2cost + P3cost	P1cost + P2cost + P3cost	P1cost + P2cost + P3cost
Items purchased								P1, P2, 1xP3

Not all CDP1-C and CDS1-C representation structures are common. In particular, the physical representation structures (that is, screens and their components), and the abstract representation structures necessary to render them (that is, page layouts and form processes) are not the same in CDP1-C and CDS1-C. The latter are not recruited to the HCI-EDP scope. The commonalities between the computer models CDP1-C and CDS1-C are recruited to the HCI-EDP scope, as shown in Figure 10.1.

The CDP1 and CDS1 computer representation states have some commonalities, shown in Table 10.2. The remaining user representation states are not common, and so are not recruited to the EDP 1 scope.

158 10. INITIAL HCI ENGINEERING DESIGN PRINCIPLES FOR B-TO-CONSUMER

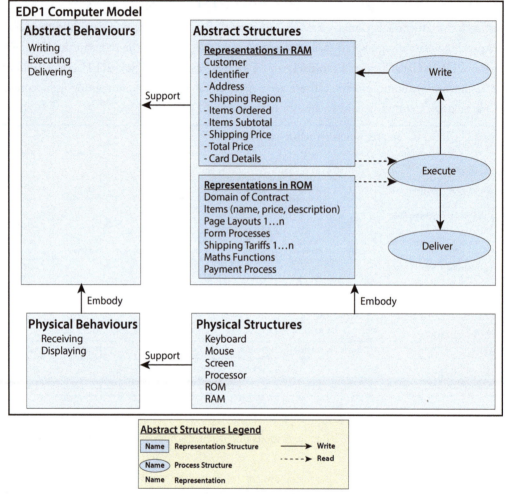

Figure 10.1: EDP1 Computer Model (EDP1-C) (following Cummaford, 2007).

Table 10.2: EDP1 scope: Computer representation structure states matrix							
	Start	After T1	After T2	After T3	After T4	After T5	After T6	After T7
Abstract Structure								
Representations in RAM								
Customer Identifier	?	Known	Known	Known	Known	Known	Known	Known
Customer Items ordered	?	P1	P1, P2	P1, P2, 2xP3	P1, P2, 2xP3	P1, P2, 1xP3	P1, P2, 1xP3	P1, P2, 1xP3
Customer Items subtotal	0	P1 cost	P1cost + P2cost	P1cost + P2cost + 2xP3cost	P1cost + P2cost + 2xP3cost	P1cost + P2cost + P3cost	P1cost + P2cost + P3cost	P1cost + P2cost + P3cost
Customer card details	?	?	?	?	?	?	?	Known

10.2 HCI ENGINEERING DESIGN PRINCIPLES ACQUIRED IN CYCLE 1 DEVELOPMENT 159

The remaining components of CDP1 and CDS1 are not identical, and so are not recruited to the EDP1 scope.

10.2.2 HCI ENGINEERING DESIGN PRINCIPLE SPECIFICATION

The HCI-EDP prescriptive design knowledge is synthesised from the non-common aspects of CDS1 and CDP1. The relevant components of CDS1 and CDP1 are first compared to identify their non-commonalities. The latter are then used to construct the prescriptive component of EDP1. User and computer representation structure states are compared first, then user and computer behaviours. The computer physical structures, sufficient to support these user and computer behaviours, are referenced to the user and computer behaviours.

Identification of CDS-only and CDP-only Components

The abstract structures comprise user representation structure states and computer representation structure states.

User Representation Structure States

The CDS1 and CDP1 user representation structure states have some commonalities. They are represented in the EDP1 Scope. The remaining CDS1 and CDP1 user representation states are not common, and so are included here, see Tables 10.3 and 10.4.

Table 10.3: CDS1 user representation states: Non-commonalities

	Start	After T1	After T2	After T3	After T4	After T5	After T6	After T7
Plan for shopping								
Shipping price	£0	Shipcost	Shipcost	Shipcost	Shipcost	Shipcost	Shipcost	Shipcost
Total price	£0	P1 cost + Shipcost	P1 cost + P2 cost + Shipcost	P1 cost + P2 cost + 2xP3 cost + Shipcost	P1 cost + P2 cost + 2xP3 cost + Shipcost	P1 cost + P2 cost + P3 cost + Shipcost	P1 cost + P2 cost + P3 cost + Shipcost	P1, cost + P2 cost + P3 cost + Shipcost,

Table 10.4: CDP1 user representation states: Non-commonalities

	Start	After T1	After T2	After T3	After T4	After T5	After T6	After T7
Plan for shopping								
Shipping price	£0	?	?	?	Shipcost	?	Shipcost	Shipcost
Total price	£0	?	?	?	P1 cost + P2 cost + 2xP3 cost + Shipcost	?t	P1 cost + P2 cost + P3 cost + Shipcost	P1, cost + P2 cost + P3 cost + Shipcost,

Computer Representation Structure States

The CDS1 and CDP1 computer representation states have some commonalities. These are represented in the EDP1 Scope. The remaining CDS1 and CDP1 computer representation states are not common, and so are included here (see Tables 10.5 and 10.6).

Table 10.5: CDS1 computer representation states: Non-commonalities

	Start	After T1	After T2	After T3	After T4	After T5	After T6	After T7
Abstract Structure								
Representations in RAM								
Customer address	?	?	?	?	?	?	?	Known
Customer shipping region	?	Known	Known	Known	Known	Known	Known	Known
Customer shipping price	0	Shipcost	Shipcost	Shipcost	Shipcost	Shipcost	Shipcost	Shipcost
Customer total price	0	P1 cost + Shipcost	P1 cost + P2 cost + Shipcost	P1 cost + P2 cost + 2xP3 cost + Shipcost	P1 cost + P2 cost + 2xP3 cost + Shipcost	P1 cost + P2 cost + P3 cost + Shipcost	P1 cost + P2 cost + P3 cost + Shipcost	P1, cost + P2 cost + P3 cost + Shipcost,

Table 10.6: CDP1 computer representation states: Non-commonalities

	Start	After T1	After T2	After T3	After T4	After T5	After T6	After T7
Abstract Structure								
Representations in RAM								
Customer address	?	?	?	?	Known	Known	Known	Known
Customer shipping region	?	?	?	?	Known	Known	Known	Known
Customer shipping price	0	?	?	?	Shipcost	Shipcost	Shipcost	Shipcost
Customer total price	0	?	?	?	P1 cost + P2 cost + 2xP3 cost + Shipcost	P1 cost + P2 cost + P3 cost + Shipcost	P1 cost + P2 cost + P3 cost + Shipcost	P1, cost + P2 cost + P3 cost + Shipcost,

CDS-only and CDP-only Physical Structures

The computer physical structures, which support the user behaviours, specified in the task goal structure, are embodied in the screens, specified for CDS1 and CDP1. The latters' task goal structures are each analysed by task. These computer physical structures are specified for each screen. The former, which are not also present in the CDP, are then identified. The computer physical structures, which are present in the CDS, but not in the CDP, are carried forward to the HCI-EDP prescriptive component.

The computer physical structures for Screen 9, referenced in Row 9 of CDS-TGS T3, are shown in Figure 10.2 by way of illustration.

Required computer structures to support user behaviours (on the related screen) are: Item Quantity Box; Add to Cart Button; Last Item Added; Item Subtotal; and Ship Cost and Total Price.

The method specified earlier is now used to identify those aspects of the CDP not represented in the CDP-CDS commonalities. While the CDP-only components are not candidates for inclusion in the HCI-EDP, they may contribute to its specification by negation. That is, if the CDP-only components are x, the HCI-EDP should specify not x. The CDP-only components are not exemplified here, but include computer structures such as a Recalculate Button on the Shopping Cart Page.

10.2 HCI ENGINEERING DESIGN PRINCIPLES ACQUIRED IN CYCLE 1 DEVELOPMENT

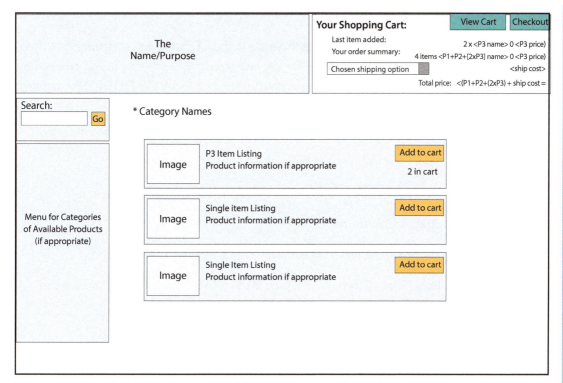

Figure 10.2: CDS1-C S9: 2x item 3 added (following Cummaford, 2007).

The CDS-only and the negation of the CDP-only components are now synthesised to construct the HCI-EDP prescriptive component, comprising user and computer representation structure states, and supported user and computer behaviours.

User Representation Structure States

The user should be aware of the shipping cost and total price throughout their interaction with the e-shop. The user representation structure states are then identical to those shown in Table 10.4.

Computer Representation Structure States

The computer should elicit sufficient information to enable the total price to be known throughout the transaction, that is, sufficient to calculate the shipping costs. The prescribed computer representation structure states are identical to those shown in Table 10.5.

Computer Physical Structures

The section comprises CDS-only components and CDP-only components.

CDS-only Components

The CDS screens are sufficient to support the behaviours in the CDS task goal structure. The CDS1 computer structures, not present in the CDP1 computer structures, are listed below, grouped by CDS1 screen. These non-commonalities are necessary components of the HCI-EDP computer structures screens.

Screen 2: Last Item Added to Cart; Goods Subtotal, and Shipping Option Selector.

Screen 3: Ship Cost and Total Price.

Screen 5: Last Item Added, Item Subtotal, Ship Cost, and Total Price.

Screen 7: Add to Cart Button.

Screen 8: Last Item Added, Item Subtotal, Ship Cost, and Total Price.

Screen 9: Item Quantity Box, Add to Cart Button, Last Item Added, Item Subtotal, Ship Cost, and Total Price.

Screen 10: Up and down arrows to change quantity.

Screen 11: Ship Cost and Total Price.

Screen 13: Total Price.

Screen 15: Items in Order and Total Price.

CDP-Only Components

The following structures are present in the CDP1 computer structures, but not in the CDS1 computer structures.

Screen 3: Change Quantity Textbox and Recalculate Button.

Screen 7.1a: Shipping Times and Prices Link.

Screen 8.2a: International Shipping Rates Link.

Screen 8.3a: International Shipping Tariffs.

The structures in Screen 3 are not required in the CDS screens, as the item quantities in the shopping basket screen are updated by the Up and Down Arrow buttons.

The shipping times and prices information, and the links thereto, are not required in the CDS screens. The shipping costs are calculated and displayed throughout the transaction.

10.3 HCI ENGINEERING DESIGN PRINCIPLES ACQUIRED IN CYCLE 2 DEVELOPMENT   163

Screens

The CDS1 screens are recruited to the EDP1 screens. The CDS1-only non-commonalities identified earlier are the necessary components to support the EDP1-TGS, and the EDP1 screens are sufficient to achieve performance Pa = Pd.

Task-Goal Structure

CDS1-TGS is recruited to form the HCI-EDP-TGS.

10.2.3 HCI ENGINEERING DESIGN PRINCIPLE ACHIEVABLE PERFORMANCE

The class design solution actual performance, as task quality and worksystem costs, is recruited from the class design solution to form the HCI-EDP class of achievable performance. The EDP1 achievable performance is shown in Table 10.7.

Table 10.7: CDS1 cost matrix (CDS1-CM)									
CDS1 (SDS1a + SCS1b)/2									
		T1	T2	T3	T4	T5	T6	T7	Totals
Absract behaviours	Plan	5	3	3	2	4	2	3	22
	Control	4	5	5	0	3	0	12	29
	Encode	4	2	3	1	3	1	5	19
	Execute	2	2	2	9	2	0	10	18
	Total	**15**	**12**	**13**	**3**	**12**	**3**	**30**	**88**
Physical behaviors	Search	3	2	3	0	3	0	5	16
	Click	3	3	3	0	2	0	27	38
	Keystroke	0	10	0	0	0	0	107	115
	Total	**6**	**13**	**6**	**0**	**5**	**0**	**139**	**169**
Task completion	**Percentage**	**100.0%**	**100.0%**	**100.0%**	**100.0%**	**100.0%**	**100.0%**	**100.0%**	**100.0%**
Time to complete (secs)	**Average**	**10.5**	**16.333**	**20.083**	**19.583**	**13.583**	**5.0833**	**57.417**	**142.583333**

This completes the acquisition of EDP 1 from Cycle 1 development.

10.3 HCI ENGINEERING DESIGN PRINCIPLES ACQUIRED IN CYCLE 2 DEVELOPMENT

10.3.1 HCI ENGINEERING DESIGN PRINCIPLE SCOPE

CDP2-D and CDS2-D are identical in terms of objects and their attributes. The EDP2 domain model (see Figure 10.3), termed EDP2-D, is thus identical to both CDP2-D and CDS2-D.

Domain Model

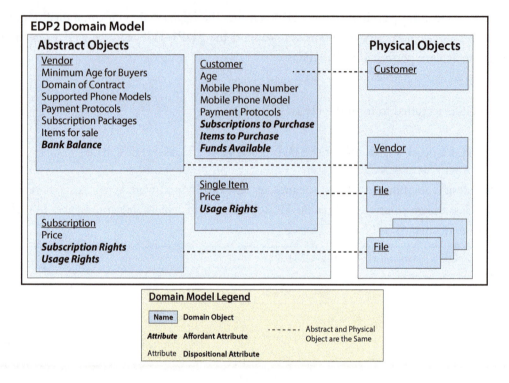

Figure 10.3: EDP2 Domain Model (EDP2-D) (following Cummaford, 2007)

Product Goal

CDP2-PG and CDS2-PG, and their pre-purchase requirements, are identical and so are recruited to the EDP scope. The EDP2 pre-purchase requirements, termed EDP2-PG are specified in Table 10.8.

10.3 HCI ENGINEERING DESIGN PRINCIPLES ACQUIRED IN CYCLE 2 DEVELOPMENT

Table 10.8: EDS2-PG: affordant object attribute value transformations

Domain Object: Attribute [value]	Start State	End State
Customer: items to purchase [{set}]	Set	
Item: usage rights[value]		Customer
Customer: subscriptions to purchase [{set}]	Set	
Premium Subscription: usage rights [value]		Customer
Customer: funds available [amount]	Amount	Amount minus total price [amount]
Vendor: bank balance [amount]	Amount	Amount plus total price [amount]

CDP2-D and CDS2-D are identical in terms of objects and their attributes. The EDP2 domain model (see Figure 10.4), termed EDP2-D, is then identical to both CDP2-D and CDS2-D.

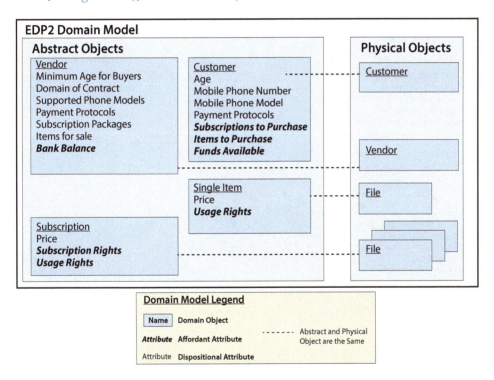

Figure 10.4: EDP2 domain model (EDP-D) (following Cummaford, 2007).

CDP2-PG and CDS2-PG, and their pre-purchase requirements, are identical, and so are recruited to the EDP scope. The EDP2 pre-purchase requirements, termed EDP2-PG are specified in Table 10.9.

166   10. INITIAL HCI ENGINEERING DESIGN PRINCIPLES FOR B-TO-CONSUMER

Table 10.9: EDP2-PG: affordant object attribute value transformations

Domain Object: Attribute [value]	Start State	End State
Customer: items to purchase [{set}]	Set	
Item: usage rights		Customer
Customer: subscriptions to purchase [{set}]	Set	
Premium Subscription: usage rights [value]		Customer
Customer: funds available [amount]	Amount	Amount minus total price [amount]
Vendor: bank balance [amount]	Amount	Amount plus total price [amount]

User Model

CDP2-U and CDS2-U are identical, and so are recruited to the EDP scope.

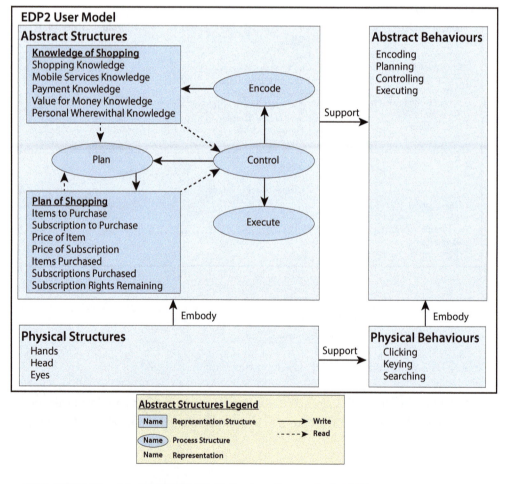

Figure 10.5: EDP2 User Model (EDP2-U) (following Cummaford, 2007).

10.3.2 HCI ENGINEERING DESIGN PRINCIPLE SPECIFICATION

Computer Physical Structures

The required computer physical structures to support user behaviours are shown in Figure 10.6.

Figure 10.6: CDS2 screen S4.2 confirmation page (Subscription) (following Cummaford, 2007).

The required computer physical structures to support user behaviours are shown in Figure 10.7.

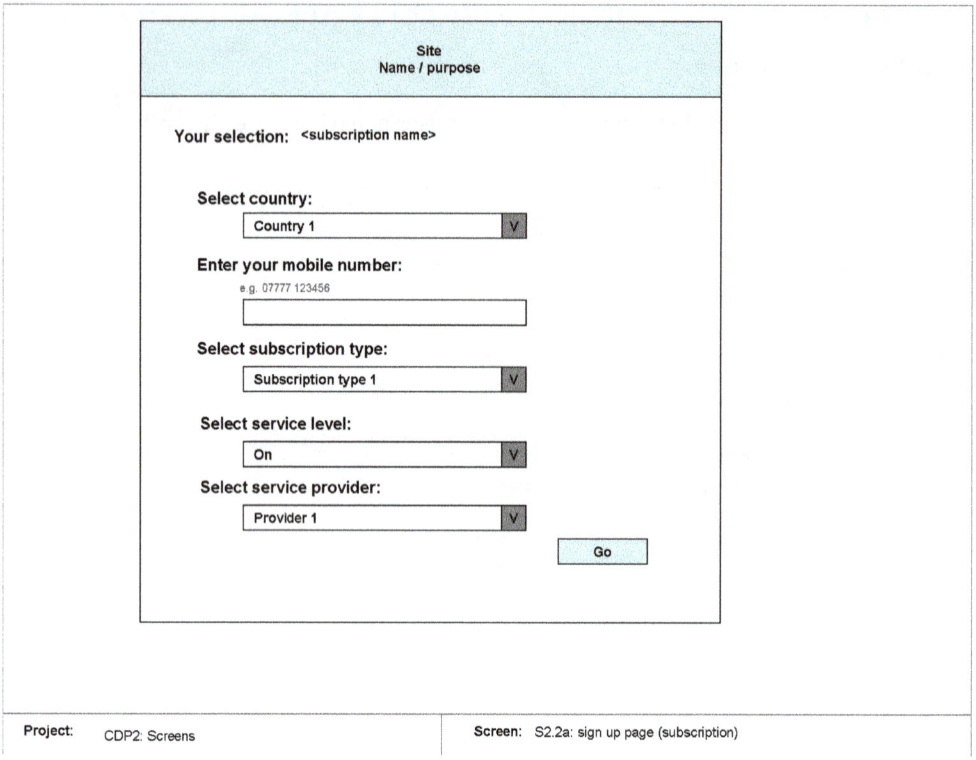

168 10. INITIAL HCI ENGINEERING DESIGN PRINCIPLES FOR B-TO-CONSUMER

Project: CDP2: Screens	**Screen:** S2.2a: sign up page (subscription)

Figure 10.7: CDP2 Screen S2.2a Sign Up Page (Subscription) (following Cummaford, 2007)

The required computer physical structures to support user behaviours are shown in Figure 10.8.

10.3 HCI ENGINEERING DESIGN PRINCIPLES ACQUIRED IN CYCLE 2 DEVELOPMENT

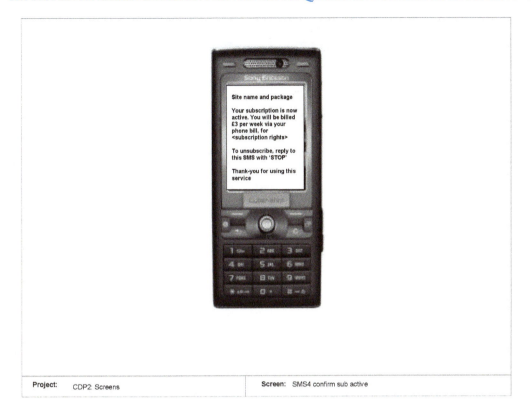

Figure 10.8: CDP2 Screen SMS4: Confirm Subscription Active (following Cummaford, 2007)

Task Goal Structures

The task-goal structures are shown in Table 10.10 CDS2 Task Goal Structure (TGS) Task 5a.

170   10. INITIAL HCI ENGINEERING DESIGN PRINCIPLES FOR B-TO-CONSUMER

Table 10.10: CDS2 task goal structure (TGS) Task 5a

R	U/C	Task description	Dom trans.	UB(abstract)	UB(physical)	CB (abstract)	CB (physical)
1	C	Display homepage [S1]				Execute: form process. Deliver: S1	Display: sign up link
2	U	Search for sign up link and select		Encode page. Control x2: get next item to cancel, choose item. Execute action	Search screen. Click on item.		
3	C	Display sign up page [S2.2a]				Execute: form process. Deliver: S2.2a	Receive: show sign up page command. Display: country dropdown, mobile number entry box, service dropdown, service level dropdown, send button
4	U	Read page, Enter details		Encode: read page, control x3: recall mobile number, problem solve - add country code, recall service to sign up for. Execute x 2: write phone number, execute action	Search screen. Click country dropdown, select 'UK residents' (2 clicks). Enter mobile number (click box, 11 keystrokes), select service (click box, select), select service level (click box, select), select operator (click box, select), click send		
5	C	Display validation code entry page [S3.2], send validation code SMS [SMS1]				Write: customer identifier, customer mobile no, customer subscription package. Execute: form process. Deliver: S3.2. Execute form process (SMS). Deliver SMS1	Receive: send validation code command, show validation code input page command. Display: validation code entry box. SMS: display validation code.
6	U	Enter validation code		Encode: read page. Encode: read validation code in SMS, control: check code entered correctly. Execute x2: action, write code	Search screen, search mobile screen. Type validation code (7 keystrokes), click 'next'		
7	C	Display confirmation page [S4.2]				Write: customer remaining subscription rights. Execute: form process. Deliver: S4.2.	Receive validation complete confirmation. Display: confirmation text
8	U	Confirm subscription status	Change Subscription: usage rights from[user, ongoing] to [user, until end of sub period]	Encode page. Plan: update 'subscriptions to purchase' 'subscriptions ordered'. Control: understand that subscription task completed	search screen.		

User Representation Structure States

The user should be aware of the price of single items and subscriptions throughout their interaction with the e-shop. Hence, the user representation structure states are as shown in Table 10.11.

Table 10.11: EDP2 prescribed user representation structure states						
	Start	After T1	After T2	After T3	After T4	After T5
Plan for shopping						
Price of item	I1 price	I1 price	I1 price	I1 price	I1 price	I1 price
Price of subscription	S1 price	S1 price	S1 price	S1 price	S1 price	S1 price

10.3.3 HCI ENGINEERING DESIGN PRINCIPLE ACHIEVABLE PERFORMANCE

The EDP2 achievable performance is shown in Table 10.12, for the tasks specified in EDP2-TGS.

Table 10.12: EDP2 costs matrix (EDP2-CM)							
CDS1 (SDS1a + SCS1b)/2							
		T1	T2	T3	T4	T5	Totals
Absract behaviours	Plan	2	1	2	1	2	8
	Control	5	0	7	0	4	16
	Encode	5	1	5	1	3	15
	Execute	5	0	5	0	2	12
	Total	**17**	**2**	**19**	**2**	**11**	**51**
Physical behaviors	Search	4	1	5	1	3	14
	Click	5	0	7	0	4	16
	Keystroke	18	0	24	0	17	59
	Total	**27**	**1**	**36**	**1**	**24**	**89**
Task completion	**Percentage**	**100.0%**	**100.0%**	**100.0%**	**100.0%**	**100.0%**	**100.0%**
Time to complete (secs)	**Average**	**46.7**	**1.6**	**33.1**	**1.9**	**27.3**	**110.5**

This completes the acquisition of EDP 2 from Cycle 2 development.

10.4 INITIAL HCI ENGINEERING DESIGN PRINCIPLES

The initial HCI-EDPs offer promise for the specification of a "parent" of such principles via abstraction. However, this was not achieved during the timescale of the research. The informal similarities, which indicate what might be specified in an initial engineering design problem at the parent-class level, are twofold.

First, the price of items should be clearly displayed throughout the transaction.

Second, the total price payable should be made available to the user as early as possible in the transaction. Any user costs incurred, in providing the minimal data required to specify the total price, are to be considered acceptable. For example, the user costs involved in entering the shipping

172 10. INITIAL HCI ENGINEERING DESIGN PRINCIPLES FOR B-TO-CONSUMER

option in CDS1 would be acceptable, as the shipping option data enable the total price to be known earlier.

The specification of an initial engineering design problem at the parent-class level would enable its application to solve design problems, which are instances of the parent class, but not of either of the two subclasses identified. However, such application would support better specification of the scope of each of the classes. The identification of design knowledge at an appropriate level of generality is a key aspect for the development of performance guarantees.

In conclusion, the HCI engineering design problem is specified. The method, reported and exemplified, is operationalised for the class design problems and class design solutions of Cycle 1 and Cycle 2 to acquire initial EDP 1 and EDP 2.

REVIEW

The chapter illustrates Cycle 1 and Cycle 2 acquisition of initial HCI-EDPs for business-to-consumer electronic commerce and the application of the associated specification method. The chapter includes: the scope of the HCI-EDP; its prescriptive design knowledge; the synthesis of its prescriptive component, and its achievable performance. Initial, early HCI-EDPs are identified.

10.5 PRACTICE ASSIGNMENT

10.5.1 GENERAL

Read § 10.1, concerning the HCI-EDP requirements.

- Check the requirements informally for completeness and coherence, as required by the case study of business-to-consumer electronic commerce.

- The aim of the practice assignment is for you to become sufficiently familiar with the Cycle 1 HCI-EDP requirements to apply it subsequently, and as appropriate, to a different domain of application.

Hints and Tips

Difficult to get started?

Re-read the assignment task carefully.

- Make written notes and in particular list the sections, while re-reading § 10.1

- Think about how the sections might be applied to describe the HCI engineering design principle specification of a novel domain of application.

- Re-attempt the assignment.

Test

List the sections from memory.

Read § 10.2 concerning the Cycle 1 HCI-EDP acquisition.

- Complete as for the previous section, starting Read § 10.1.

Read § 10.3 concerning the Cycle 2 HCI-EDP acquisition.

- Complete as for the previous sections, starting Read § 10.1 and Read § 10.2.

10.5.2 PRACTICE SCENARIO

Practice Scenario 10.1: Applying the Acquisition of HCI Engineering Design Principles to an Additional Domain of Application

Select a domain of application, different from that of business-to-consumer electronic commerce.

- Apply the HCI-EDP acquisition cycle for business-to-consumer electronic commerce (§ 10.1–3) to the novel domain of application. The latter can only be of the most general kind - that is at the level of the sections. However, even consideration at this high level can orient the researcher towards acquisition of the HCI-EDPs to novel domains. The latter are as might be required subsequently by their own work. The practice scenario is intended to help bridge this gap.

CHAPTER 11

Assessment and Discussion of Initial HCI Engineering Design Principles for Business-to-Consumer Electronic Commerce

SUMMARY

This chapter assesses and discusses the acquisition of initial HCI-EDPs for the domain of business-to-consumer electronic commerce. The need for further research is identified.

11.1 INTRODUCTION

The challenge for HCI research is the lack of validated, and so effective, HCI design knowledge with guarantees of application to design practice. The response is the acquisition of class-level HCI design knowledge. The initial HCI-EDPs acquired have yet to be validated. However, they are claimed to be validatable and so potentially more effective.

Issues arising from operationalisation of the strategy to acquire HCI-EDPs are identified to support future research.

11.2 STRATEGY ASSESSMENT

Operationalisation of the "initial class" strategy acquired initial HCI-EDPs. It offers, then, promise of design practice supported by performance guarantees. The specification of class design problems and solutions is viable. The "class first" strategy is considered successful. The latter includes extending the conception of HCI-EDPs, specifying a strategy and methods for their acquisition and conducting an initial operationalisation of the strategy by two cycles of development.

11.3 DISCUSSION

A number of issues are identified for future research. Their presentation follows the stages of the operationalisation method.

11.3.1 SPECIFIC DESIGN PROBLEMS SPECIFICATION

The section comprises selection of: systems; tasks; specific design problem prototypes; empirical testing; and models, development, and specification.

11.3.1.1 Systems for Specific Design Problem Development Selection

Only two systems are tested, to develop each class design problem. Testing more systems would give greater confidence in abstracted similarities between instances.

Cycle 1 systems are very similar, in terms of computer structures and behaviours. The latter are appropriate to ensure specification of CDP1. Cycle 2 systems are less similar, in terms of: computer structures and behaviours; task goal structures, and product goal. These differences support a more comprehensive evaluation of the class design problem/class design solution specification method.

11.3.1.2 Task Selection

Task goal structure tasks include typical shopping activities. The latter characterise transactions performed on e-commerce sites. The tasks are standardised to support systematic comparison of the systems, with respect to each other and to the prototypes. Future research needs to consider sub-task frequency.

For e-commerce, the selection of goods, reception of the collated order (for physical goods) and payment are necessary tasks. The product goals are defined to ensure task completion to achieve the product goal. Future research needs to address task selection, based on the frequency and occurrence of tasks using existing systems. Other domains, for example, air traffic management, have a commonly agreed corpus of tasks. Development of HCI-EDPs for such domains would avoid issues of task selection completeness.

Defining tasks with specific ordering and products enables systematic comparison of performance between users across tasks. Users performing tasks in any order could be studied, if the start and end points of each are defined.

11.3.1.3 Specific Design Problem Prototypes

While actual payment was simulated, users reported they would be uncomfortable entering payment authorisation, before knowing the total price. Although Tq is based on a simulation, testing the latter is considered successful. Users' subjective ratings of workload provide confirmation. However, using functioning e-shops to measure performance would produce more detailed data.

11.3.1.4   Empirical Testing

The class design problem testing was conducted after development of their respective class design solutions. The former could not, therefore, have informed the latter. Further testing was carried out, but is not reported. Testing order should be reviewed

The sample size of six participants per specific design problem/specific design solution test was determined by the timescale of the research. Larger numbers of participants would increase confidence in the results.

11.3.1.5   Model Selection, Development, and Specification

The user, computer and domain models are based on research by Hill (2010), which applies the Dowell and Long conception (1989). They perforce contain the elements to operationalise the conception. Alternative representations for user, computer and domain models should be considered.

The process of model development per se has not been reported. It is similar for each specific design problem. The summary here is to support future research. The domain transformations to achieve the product goal are specified first, to inform development of the domain model. This ensures that it contains only those objects, whose attribute values are transformed by the work. The task goal structure is then specified. First, the user and computer behaviours sufficient to achieve the domain transformations, specified in the product goal, are identified. Second, the user and computer structures, sufficient to support the behaviours identified first are specified. The user and computer models are then specified, by identification of the minimal set of user and computer structures, sufficient to support the behaviours identified in the task goal structure.

Cycle 1 iteration of model development does not include user and computer structures in the task goal structure. The lack, then, excludes the explicit relation of the latter to the former. So the models require iteration. Resulting user and computer model insights inform the reverse engineering of the task goal structure, domain model and product goal.

User behaviours are assumed equivalent for workload (Uc). They involve user workload, but the relative workload of each may not be equal. Empirical evidence for differential workload between behaviours could be integrated into the costs matrix. Physical costs are enumerated by counting observable user behaviours, but abstract costs are inferred from user task performance and verbal protocols. Standardised criteria inform identification of user abstract costs. The latter assume page-encoding costs to be equivalent to that of mental (structure) activation. Future research should explore the criteria, and costs equivalence, for behaviours and activation of associated structures.

11.3.2 CLASS DESIGN PROBLEM SPECIFICATION

Pd for the class design problem is specified relative to the Pa of the specific design solutions tested (that is, increase Tq, reduce Uc). It is possible, in principle, to state Pd in absolute values. Future research could address this issue.

User abstract and physical behaviour counts are not included in the class design problem task goal structure tables, even when the latter contain a table from the specific design problem. It was decided to remove the behaviour counts, because the class design problem costs matrix contains the mean count for each user behaviour, for all specific and class design problems.

11.3.3 CLASS DESIGN PROBLEM EVALUATION

Analytic evaluation of the class design problem involves checking that sufficient structures and behaviours are present in the user and computer models, to perform the task goal structure of each specific design problem. The assessment supports evaluation of the class design problem user and computer models for the development of HCI-EDPs.

In Cycle 2, the class design problem for some task goal structures for some tasks contains both specific design problem task goal structures, that is, two instances and no abstraction. Abstracting across two similar but not identical specific design problem task goal structures requires further research. The "chunking" of multiple behaviours would enable identification of commonalities between task goal structures, which are similar but not identical.

11.3.4 CLASS DESIGN SOLUTION SPECIFICATION

User abstract and physical behaviour counts are not included in the class design solution task goal structure, even though they are generally consistent between specific design solutions. While the counts are similar in the present case, this might not be the case for all such systems. So, the counts are calculated using the mean values across specific design solutions. This enables greater levels of generality in future design instances.

The specification of discrete user mental structures is based on the user mental behaviours identified by task analysis. Identification of separate behaviour types (for example, calculate, read, etc) is, then, an important factor impacting the final enumeration of user costs, and the contents of the user model. The process should be investigated further.

11.3.5 SPECIFIC DESIGN PROBLEM SPECIFICATION

The specific design problem prototypes exhibit features of the systems tested. Inclusion ensures that performance differences between the specific design problems and solutions are due to those of the class design problem and solution and not to specific features of the e-shops and prototypes.

The features ported from the specific design problem to its solution are not present, then, in the initial HCI-EDP. The initial such principles could not, therefore, support specification of a novel specific design solution for which no specific design problem system exists. The issue of completeness needs further investigation.

11.3.6 CLASS DESIGN SOLUTION EVALUATION

Structures are specified for both the user and computer models. However, workload is calculated without reference to structures. This suggests that structures may not be a necessary component. However, structures offer a more complete characterisation of the worksystem, and would be necessary for costs that are incurred, when class design problem and solution structures change (that is, when structure "set-up" costs are not zero).

11.3.7 CLASS DESIGN PROBLEM TO CLASS DESIGN SOLUTION MAPPING

While the class design problem and the class design solution cannot be tested directly, the differences in performances of their instantiations suggest the performance of the latter to be superior to that of the former. Subjective reports of workload indicate that the specific design problem prototypes incur less costs than the specific design problem systems. These workload ratings confirm the relative costs, derived from the costs matrices. They offer support for the analytically derived calculations of relative workload.

11.3.8 HCI ENGINEERING DESIGN PRINCIPLE DEFINITION METHOD

Two initial HCI-EDPs are proposed using the method specified earlier. The latter, then, offers promise for the development of further initial HCI-EDPs. The method identifies the computer structures necessary to support the user and computer behaviours, specified in the class design solution task goal structure. Validation of this relationship would be possible.

11.3.9 INITIAL HCI ENGINEERING DESIGN PRINCIPLES

The initial HCI-EDPs offer promise for the specification of an initial HCI-EDP for the parent class design problem, identified earlier by abstraction. However, this was not achieved. The informal similarities, which indicate what might be specified in an initial HCI-EDP at the parent-class level, are twofold. First, the price of items should be clearly displayed throughout the transaction. For this, the price of items must be knowable prior to purchase. Second, the total price payable should be made available to the user as early as possible in the transaction. This would include any user costs, associated with providing the minimal data to ensure the total price is considered acceptable.

The specification of an initial HCI-EDP at the parent-class level would enable its application to solve design problems, which are instances of the parent class, but not of either of the two subclasses identified. Such application would support better specification of the scope of each of the classes. The identification of design knowledge at an appropriate level of generality is a key aspect for the development of performance guarantees and, as such, is an important requirement for future research. The latter needs to address application of the HCI-EDPs to novel specific design problems within the scope of each EDP.

11.3.10 REQUIREMENT FOR VALIDATION, LEADING TO GUARANTEE REVIEW

The research specifies HCI design knowledge at the class level. The specification offers promise for validation of the design knowledge, leading to guarantee. This results from the completeness of the design solution specification. Design solutions are characterised by their domain transformations, enabling relations between the work, the worksystem and performance to be specified. Dowell and Long's (1989) conception of the HCI design problem is considered appropriate. The explicit specification of the domain model offers promise for validation, leading to performance guarantees.

The method for ascribing guarantees of application involves empirical testing to develop and ascribe guarantees. It may be feasible to ascribe guarantees from HCI-EDP testing by multiple researchers, on multiple specific design solutions. However, only in the case, that the testing yields consistent results.

11.3.11 HCI CONCEPTIONS REVIEW

This research applies the Dowell and Long (1989) conception of the HCI design problem for an engineering discipline of HCI (Long and Dowell, 1989). It is not a test of the engineering discipline conception itself. However, it supports a demonstrable operationalisation of the conception to achieve its stated purpose of "supporting the design of more effective human-computer worksystems" (Long, 2021). Additional conception components are required to develop HCI-EDPs namely the conception of the general HCI-EDP and the general design solution. However, progress has been made, which affords carry forward for future research.

11.3.12 HCI ENGINEERING DESIGN PRINCIPLE APPLICABILITY AND POTENTIAL AS DESIGN SUPPORT

HCI-EDPs are a long-term goal of HCI research. The form of their expression remains an issue. However, this research has delivered interim products, in the form of initial HCI-EDPs. Future research needs to build on this work and in particular on the expression required to support design practice.

11.3.13 FUTURE RESEARCH DISCUSSION

The HCI-EDPs are defined, but need to be applied and tested, to support their validation. Their application will allow the format of their expression to be evaluated to inform their further development.

Application of the user model to diagnose user costs is onerous. It is not currently well enough specified to support multiple practitioners achieve consistent results. The user model, and procedures for diagnosis of user costs, are only fit-for-purpose, once multiple analysts achieve consistent results for the same specific design problem. To achieve consistent results, the criteria for the diagnosis of mental behaviours needs to be further investigated.

This research identifies a parent class of e-commerce transaction systems, containing two subclasses. The feasibility of further subclasses, better to understand the specification of design knowledge at varying levels of generality, is desirable. In addition, the boundary conditions of the two subclasses needs to be investigated. While the subclass of physical goods transaction systems, identified in Cycle 1, may well include a great number of specific design problems, the class criteria should be reviewed in light of specific design problems which are similar in some respects, but may be solvable, using an alternative class design solution.

Future research will also need to take account of the changes, which have occurred in best practice, since the research was carried out. The research itself applied the best practice of the day to support the acquisition of the initial HCI-EDPs. It also raised the issue of the format of such principles best suited to design practice application.

Last, the HCI-EDP conception and definition methods should also be operationalised for other classes of design problem to investigate the generality of the methods across multiple classes of design problem.

11.4 BUSINESS-TO-CONSUMER BEST PRACTICE UPDATE

Revenue for digital media and technology in general, and e-commerce in particular, has grown. Turnover for e-commerce has increased from about £20 billion in 1998 to about £2 trillion per annum globally. Hardly surprising then, that commercial best-practice has attracted resources, resulting in its development and advancement. Future research needs to take account both of how to apply current best-practice in the acquisition of HCI-EDPs and what format best suits their application to current best-practice design. To the latter ends, the changes to best-practice, since the completion of the research, are identified and implications for best-practice and for HCI-EDP application format noted. Future research would do well to take account of both sets of implications. Such changes in best practice follow:

1. from design for usability to design for user experience (UX);

2. from design methods to design methods, enhanced by technical advances in data capture, as exemplified by UX analytic tools, such as Adobe Analytics and Content Square;

3. from simple online transaction testing to online "design funnel" testing;

4. from simple online transaction testing to online "AB" testing;

5. from structured analysis and design methods to "lean UX'" design methods;

6. from the design problem to the minimum "viable product" (MVP);

7. from process design methods to "atomic" design methods; and

8. from individual online user testing to online "scaled up" user testing.

Note that in all cases, the full range needs to be included.

All these best-practice changes can be recruited to the design practice, used in the case study, to support the acquisition and validation of HCI-EDPs. Such application, however, would necessarily require the mapping of the novel change concepts, such as "lean" and "minimum viable product" to those of the conception, such as "design problem" and "design solution" and, indeed, "HCI-EDP." The format of the latter for best practice application was an issue at the time of the research and remains an issue now.

In conclusion, as concerns e-commerce systems, it is clear, that they were a promising area of research, in terms of their potential for commercial development. The selection of physical goods e-commerce transaction systems has also proven to be an area of commercial interest and success. There is no shortage of such systems, with Amazon emerging as perhaps the best known, and possibly biggest. Information e-commerce systems, however, have almost disappeared in the form characterised in Cycle 2 development. The general class-level description of transaction systems for information may still be valid, for example, the sale of virtual goods in games or the metaverse, but SMS news alert services as such hardly exist. However, the particular example is less important than the development of the HCI-EDP conception and the class-based approach themselves.

REVIEW

This chapter assesses and discusses the acquisition of initial HCI-EDPs, supported by guarantees, for the application domain of business-to-consumer electronic commerce. The associated strategy and operationalisation are assessed as successful. The general discussion includes progress and carry forward, together with other issues requiring address by future research.

11.5 PRACTICE ASSIGNMENT

11.5.1 GENERAL

Read § 11.3.1, concerning the specification of specific design problems.

- Check the specification sections informally for completeness and coherence, as required by the case study of business-to-consumer electronic commerce.

- The aim of the assignment is for you to become sufficiently familiar with the specification of specific design problems to apply it subsequently and as appropriate to formulating your own assessment and discussion of initial HCI-EDPs for the case study of business-to-consumer electronic commerce in Practice Scenarios 11.1–2.

Hints and Tips

Difficult to get started?

    Re-read the assignment task carefully.

- Make written notes and in particular list the sections, while re-reading § 11.3.1.

- Think about how the sections might be applied to formulating your own assessment and discussion of initial principles for the case study of business-to-consumer electronic commerce.

- Re-attempt the assignment.

Test

List from memory as many of the sections as you can.

    Read § 11.3.2, concerning the specification of class design problems.

- Complete as for the previous practice assignment, starting Read § 11.3.1, on specific design problems.

11.5.2 PRACTICE SCENARIOS

Practice Scenario 11.1: Formulate and Apply Your Own Strategy Assessment and Discussion of the Specification of Specific Design Problems

Formulate your own strategy assessment and discussion sections to replace, as required, those of § 11.3.1. Use the latter as a basis, for your formulation, along with examples from the HCI research literature.

- Propose your own assessment and discussion for the initial HCI-EDPs for the case study of business-to-consumer electronic commerce. Your assessment may be in agreement with that offered by § 11.3.1. However, if you disagree with the latter in any way, try to be critical and to advance your own different assessment. The proposal can only be of the most general kind—that is at the level of the section. However, even application at this high level can provide practice for the reader in assessing and discussing the results of their own work.

Practice Scenario 11.2: Formulate and Apply Your Own Strategy Assessment and Discussion of the Specification of Class Design Problems

Complete as for the § 11.1 Practice Scenario on specific design problems.

CHAPTER 12

Progress in Carry Forward of HCI Engineering Design Principles for Future Research

SUMMARY

This chapter introduces the progress, in terms of carry forward, towards the acquisition of HCI-EDPs, common to the two case studies of domestic energy planning and control and of business-to-consumer electronic commerce. Future research necessary to acquire and to validate HCI-EDPs is identified.

12.1 TOWARD HCI ENGINEERING DESIGN PRINCIPLES - GENERAL PROGRESS AND CARRY FORWARD

Both research case studies claim to have made progress towards HCI-EDPs. These claims are documented for the application domain of domestic energy planning and control (Chapters 2–6) and for that of business-to-consumer electronic commerce (Chapters 7–11). Here, the claims are brought together to establish which are common to both case studies and so both domains of application, such that they can be considered general. The listing is to identify research, which can be carried forward, in the quest for HCI-EDPs. Also, to identify the associated need for future research. The listing is to facilitate carry forward by researchers in their own work.

12.1.1 DOMESTIC ENERGY PLANNING AND CONTROL

The complete set of claims is listed, as concerns the progress towards HCI-EDPs, made for the application domain of domestic energy planning and control. Associated references cite earlier and more detailed documentation:

1. Conception of substantive HCI-EDPs (§ 2.1).

2. Conceptions of the general design problem and solution (§ 2.1).

3. Conceptions of the specific design problem and solution (§ 2.1).

4. Strategy for developing HCI-EDPs (§ 2.2).

5. Conception of human-computer systems (§ 2.3).

6. Operationalisation of specific design problems and solutions (§ 3.1).

7. Conception of planning and control (§ 3.2).

8. Best-practice development (§ 3.3 and § 4.1).

9. Evaluation (§ 3.3).

10. Development cycle operationalisations (§ 3.4 and § 4.2).

11. Initial HCI-EDPs (§ 5.2, § 5.4–7).

12. MUSE for Research (MUSE/R) (§ 6.2).

12.1.2 BUSINESS-TO-CONSUMER ELECTRONIC COMMERCE

The complete set of claims is listed, as concerns the progress towards HCI engineering design principles, made for the application domain of business-to-consumer electronic commerce. Associated references cite earlier and more detailed documentation.

1. Conception of HCI-EDPs (§ 7.1).

2. Conceptions of the general design problem and solution (§ 7.2–3).

3. Specifications of specific design problems and solutions (§ 7.3).

4. Strategy for developing HCI-EDPs (§ 7.2).

5. Conception of classes of design problem and solution (§ 7.2–4).

6. Method for operationalising class-first strategy (§ 7.3).

7. Method for specifying class design problem and class design solution (§ 7.3).

8. Method for specifying HCI-EDPs (§ 7.3).

9. Identification of class design problems (§ 7.4).

10. Testing procedure (§ 8.1).

11. Development cycle operationalisations (§ 8.1 and § 10.1).

12. Initial HCI-EDPs (§ 7.1 and § 10.2).

12.1.3 RESEARCH PROGRESS

The claims that follow are common, and so general, to the two case studies and the associated application domains of domestic energy planning and control and business-to-consumer electronic commerce. Numbers refer to the domains, respectively. A single number indicates identical ordering.

1. Conception of HCI-EDPs (of some type) (1).

2. Conceptions of the general design problem and solution (2).

3. Conceptions/specifications of the specific design problem and solution (3).

4. Strategy for developing HCI-EDPs(4).

5. Operationalisation/specifications of specific design problems and solutions (6).

6. Development cycle operationalisations (10 and 11).

7. Evaluation/testing procedure (9 and 10).

8. Initial engineering HCI-EDPs (11 and 12).

The claims common, and so general, to the two case studies, associated with the domains of application, demonstrate progress towards acquiring HCI-EDPs. The general progress comprises both the number of claims and their scope. Critical outcomes, available for carry forward by future HCI-EDP research, are included. The critical outcomes, ordered for such carry forward are: acquisition of initial HCI-EDPs (of some type); strategy for the development of such the acquisition of such principles; cycle operationalisations to implement such a strategy; conceptions/specifications of general and specific design problems and solutions to support such operationalisations; and evaluation/testing to assess the solution of those general and specific problems. The case of procedural principles is addressed later, when discussing the differences between the two domains of application

12.2 HCI ENGINEERING DESIGN PRINCIPLES: RESEARCH REMAINING

Both research case studies propose future research to acquire and to validate HCI-EDPs. These proposals are documented for the application domain of domestic energy planning and control (Chapter 5) and for that of business-to-consumer electronic commerce (Chapter 10). Here the proposals are listed to establish which are common to both domains of applications, such that they can be carried forward by future research.

12.2.1 DOMESTIC ENERGY PLANNING AND CONTROL

The complete set of proposals is listed, as concerns the remaining research to acquire and to validate HCI-EDPs, made for the application domain of domestic energy planning and control. Associated references cite earlier and more detailed documentation.

1. To develop further cycles of (declarative and procedural) initial HCI-EDPs (§ 6.1).

2. To acquire HCI-EDPs (§ 6.1).

3. To develop the guarantee supporting HCI-EDPs (§ 6.1).

4. To re-express HCI-EDPs in support of design practice (§ 6.1).

5. To validate HCI-EDPs (§ 6.1).

6. To develop product and process tool support for MUSE/R, including operationalisations and identification of relationships, such as consistency (§ 6.2).

12.2.2 BUSINESS-TO-CONSUMER ELECTRONIC COMMERCE

The complete set of proposals is listed, as concerns the remaining research to acquire and to validate HCI-EDPs, made for the application domain of business-to-consumer electronic commerce. Associated references cite earlier and more detailed documentation.

1. To develop further cycles of (declarative and procedural) initial HCI-EDPs (§ 11.3).

2. To acquire HCI-EDPs (§ 11.3).

3. To develop the guarantee supporting HCI-EDPs (§ 11.3).

4. To re-express HCI-EDPs in support of design practice (§ 11.3).

5. To validate HCI-EDPs (§ 11.3).

6. To develop products and processes to support the method for specifying class design problem and class design solution and the method for specifying HCI-EDPs (§ 11.3).

12.2.3 RESEARCH REMAINING

The proposals that follow for the remaining research to acquire and to validate HCI-EDPs are common, and so general, to the two case-studies and the associated application domains of domestic energy planning and control and business-to-consumer electronic commerce. Numbers refer to the domains, respectively. A single number indicates identical ordering. The critical outcomes, ordered for carry forward by future research are: 1. To develop further cycles of declarative and procedural initial EDPs (1); 2. To acquire HCI-EDPs (2); 3. To develop the guarantee supporting HCI-EDPs (3); 4. To re-express HCI-EDPs in support of design practice (4); and 5. To validate HCI-EDPs (5).

The proposals that are common, and so general, to the two domains of application demonstrate some consensus, concerning the research remaining to acquire and to validate HCI-EDPs. The general consensus comprises both the number of proposals and their scope. Critical requirements to be carried forward by future research are included. The latter, as concern HCI-EDPs and listed in dependent order, are: development cycles; acquisition; guarantee development, and re-expression to support design and validation.

REVIEW

The chapter summarises the research progress towards HCI-EDPs for the application domains of domestic energy planning and control and business-to-consumer electronic commerce. The progress, common to both case studies, is identified and so generalised. This progress constitutes possible carry forward by future research. The chapter then summarises the research remaining to acquire and to validate HCI-EDPs for both the domains of application. The research remaining, common to both case studies, is identified and so generalised.

12.3   PRACTICE ASSIGNMENT

12.3.1 GENERAL
Read § 12.1.1.

- The claims of progress are listed briefly for the purposes of generalisation over domains and to facilitate carry forward by future research. To ensure comprehension of the domestic energy planning and control concepts involved, for subsequent application to this and other domains, complete the following.

- Add details to the expression of the main concepts, as they relate to progress. The former can be at the same or at a lower level of description. The details can be found in the references to earlier documentation, presented in the listing. For example, the strategy for the domestic energy planning and control case study is originally claimed to be "bottom up," but is later claimed to acquire some "top down" aspects during the conduct of the research.

- Identify additional references, in the associated sections to more detailed documentation, to those found in the listing of the claims to have made progress.

190 12. PROGRESS IN CARRY FORWARD OF HCI ENGINEERING DESIGN PRINCIPLES

- Do you agree with the list of progress claims for domestic energy planning and control? If not, delete those claims, with which you disagree and add those claims, with which you do agree. Rationalise your changes.

Hints and Tips

Difficult to get started?

Read the assignment tasks carefully.

- Make written notes, while re-reading § 12.1.1.

- Re-attempt the assignment.

Test

Add further details to the expression of the progress claims from memory.

Read § 12.1.2.

- The claims of progress are listed briefly for the purposes of generalisation over domains and to facilitate carry forward by future research. To ensure comprehension of the business-to-consumer electronic commerce concepts involved, for subsequent application to this and other domains, complete as for Read § 12.1.1 earlier.

Read § 12.1.3.

- Do you agree with the list of progress claims considered to be common and so general to the application domains of domestic energy planning and control and business-to-consumer electronic commerce? If not, make a list, which you do consider to be common and so general. Give reasons for your selection. Complete as for Read § 12.1.1 earlier.

Read § 12.2.1.

- The proposals for remaining research are listed briefly for the purposes of generalisation over domains. To ensure comprehension of the domestic energy planning and control concepts involved, for subsequent application to this and other domains, complete the following.

- Add details to the expression of the main concepts, as they relate to remaining research. The former can be at the same or at a lower level of description. The details can be found in the references to earlier documentation in the listing. For example, the claim that HCI-EDPs require the support of a guarantee for their application to design practice.

- Identify additional references in the associated sections to more detailed documentation to those found in the listing of the claims to be remaining research.

- Do you agree with the list of the remaining research proposals for domestic energy planning and control? If not, delete those proposals, with which you disagree and add those proposals, with which you do agree. Rationalise your changes. Complete as for Read § 12.1.1 earlier.

Read § 12.2.2.

- The proposals for research remaining are listed briefly for the purposes of generalisation over domains and to facilitate carry forward by future research. To ensure comprehension of the business-to-consumer electronic commerce concepts involved, for subsequent application to this and other domains, complete the following.

- Add details to the expression of the main concepts, as they relate to research remaining. The former can be at the same or at a lower level of description. The details can be found in the references to earlier documentation, to be found in the listing. For example, it is claimed that HCI-EDPs require the support of a guarantee.

- Identify additional references in the associated sections to more detailed documentation to those found in the listing of the proposals for research remaining.

- Do you agree with the list of the research remaining proposals for business-to-consumer electronic commerce? If not, delete those proposals, with which you disagree and add those proposals, with which you do agree. Rationalise your changes. Complete as for Read § 12.1.1 earlier.

- Add further details to the expression of the proposals for research remaining from memory.

Read § 12.2.3.

- Do you agree with the list of proposals for remaining research, considered to be common, and so general, to the application domains of domestic energy planning and control and business-to-consumer electronic commerce? If not, make a list, which you do agree to be common and so general. Give reasons for your selection. Complete as for § 12.1.1 earlier.

192 12. PROGRESS IN CARRY FORWARD OF HCI ENGINEERING DESIGN PRINCIPLES

12.3.2 PRACTICE SCENARIOS

Practice Scenario 12.1: Applying Your General Research Progress Claims to an Additional Domain of Application

Select the common, and so general, research progress claims

- Select an additional domain of application.

- Apply the main concepts, appearing in your list, to the novel domain of application. The description can only be of the most general kind—that is at the level of the concepts in the list. However, even consideration at this high level can orient the researcher towards application of the concepts to novel domains of application. The latter are as might be required subsequently by their own work. The research design scenario is intended to help bridge this gap.

Practice Design Scenario 12.2: Applying Your General Remaining Research Proposals to an Additional Domain of Application

Select the common, and so general, remaining research proposals.

Select an additional domain of application.

- Apply the main concepts, as for Practice Scenario 12.1 earlier.

Postscript

The Preface makes clear the book's aims. It is now time to consider how well or not they have been met.

First, the scope and content of HCI as a discipline, HCI as engineering, and its associated design problem have been specified by review and operationalised by case study.

Second, following a critique of HCI design knowledge, the challenge of a better guarantee of HCI design knowledge, including that of HCI as engineering, to support HCI design practice, has been identified, documented, and addressed.

Third, HCI-EDPs have been conducted in two case studies, as one way of meeting the challenge of the better guarantee required of HCI design knowledge to support more effective HCI design practice.

Fourth, HCI-EDP instance-first and class-first approaches to the acquisition of HCI-EDPs have been developed and implemented in two case studies of different domains of application.

Fifth, early (that is, incomplete) and initial (that is, unvalidated) HCI-EDPs have been acquired. Both the instance-first and class-first approaches, then, need further development. Some progress, however, has been demonstrated and it affords carry forward by future HCI-EDP research. The need for future such research, however, is extensive and is in no way underestimated here.

Sixth, further research is also required, concerning alternatives to the HCI-EDPs, proposed, to improve the reliability of HCI design knowledge.

Seventh, additional research of this range and magnitude will require researchers to build on each other's work. Such building needs to create a better consensus, concerning HCI discipline progress as the acquisition and validation of HCI design knowledge to support HCI design practice effectively.

The book's aims, then, are considered to be met, at least in part. HCI-EDPs indeed have been acquired. However, they are modest in number and mostly by example. Further, they remain incomplete and unvalidated. It is very much a "toward" book, as recognised in its title. At best, a start has been made and much ground has been cleared. Some progress has been made and affords carry forward by future research. For future researchers, however, it will be more "attempting to follow in the footsteps," rather than "standing on the shoulders." Good luck to those researchers willing to have a go at taking up the challenge!

Bibliography

Alterman, R. (1988). Adaptive planning. *Cognitive Science*, 12, 393-421. DOI: 10.1207/s15516709cog1203_3. 36

Atwood, M., Gray, W., and John, B. (1996). Project Ernestine: analytic and empirical methods applied to a real world CHI Problem. In Rudisill, M., Lewis, C., Polson, P. and McKay, T. (Ed.s), *Human Computer Interface Design: Success Stories, Emerging Methods and Real World Context.* 5

Barnard, P. (1991). Bridging between basic theories and the artifacts of human-computer interaction. Carroll, J. (Ed.), *Designing Interaction*, UK: Cambridge University Press. 3

Bayle, E. et al. (1997). Putting it all together: towards a pattern language for interaction design. *CHI'97 Workshop.* 5

Bevan, N. (2001). International standards for HCI and usability. *International Journal of Human-Computer Studies*, 55(4), 533-552. DOI: 10.1006/ijhc.2001.0483. 4

Boehm, B. and Lane, J. (2006). 21st century processes for acquiring 21st century systems of systems. *Cross-talk: the Journal of Defense Software Engineering*, 19(5), 4-9. 3

Card, S., Moran, T., and Newell, A. (1983). *The Psychology of Human-Computer Interaction.* Hillsdale, NJ: LEA. 2, 3

Carroll, J. (2003). Introduction: Toward a multidisciplinary science of human-computer interaction. In Carroll, J. (Ed.) *HCI Models, Theories and Frameworks.* San Francisco, CA: Morgan Kaufmann. DOI: 10.1016/B978-155860808-5/50001-0. 3

Carroll, J. (2010). Conceptualizing a possible discipline of human-computer interaction. *Interacting with Computers*, 22(1), 3-12. DOI: 10.1016/j.intcom.2009.11.008. 3

Colbert, M. (1994). Carry forward in the development of military planning systems. Unpublished Ph.D. thesis, University of London. 33, 36, 37, 38, 40

Cummaford, S. (2000). Validating effective design knowledge for re-use: HCI engineering design principles. In *CHI '00 Extended Abstracts on Human Factors in Computing Systems.* New York, NY: ACM Press. DOI: 10.1145/633292.633336. 4

Cummaford, S. (2007). HCI engineering design principles: Acquisition of class-level knowledge. Unpublished Ph.D. Thesis, University of London. xxii, 4, 5, 6, 101, 106, 109, 110, 111, 125, 128, 132, 134, 136, 140, 142, 144, 158, 161, 164, 165, 166, 167, 168, 169

196 BIBLIOGRAPHY

Cummaford, S. and Long, J. (1998). Towards a conception of hci engineering design principles. In *Proceedings of Ninth European Conference on Cognitive Ergonomics (ECCE9)*, Limerick, Ireland. 4, 5

Cummaford, S. and Long, J. (1999). Costs matrix: Systematic comparisons of competing design solutions. In *Proceedings INTERACT 99*, Volume II, Edinburgh UK, Aug 33–Sept 3, 1999. 100, 103

Denley, I. and Long, J. (2001). Multi-disciplinary practice in requirements engineering: problems and criteria for support. In Blandford, A., Vanderdonkt, J., and Gray, P. (Eds) *People and Computers XV – Interaction without Frontiers. Joint Proceedings of HCI 2001 and IHM 2001*. London: Springer Verlag. DOI: 10.1007/978-1-4471-0353-0_8. 3

Dowell, J. (1993). Cognitive engineering and the rationalisation of the flight strip. Unpublished Ph.D. Thesis, University College, London. 34, 35, 36

Dowell, J. (1998). Formulating the cognitive design problem of air traffic management. *International Journal of Human-Computer Studies*, 49(5), 743-766. DOI: 10.1006/ijhc.1998.0225. 3, 24, 39, 40

Dowell J. and Long J. (1989). Towards a conception for an engineering discipline of human factors. *Ergonomics*, 32(11), 1513-1535. DOI: 10.1080/00140138908966921. 1, 3, 4, 9, 10, 11, 12, 13, 14, 16, 19, 24, 25, 29, 84, 88, 91, 92, 94, 96, 108, 177, 180

Friend, J. and Jessop, W. (1969). *Local Government and Strategic Choice: An Operational Research Approach to the Processes of Public Planning*. UK: Tavistock Publications. 35

Glaser, B. and Strauss, A. (1967). *Discovery of Grounded Theory*. London: Aldine. 3

Hallam-Baker, P. (1996). User interface requirements for sale of goods. World Wide Web Consortium Technical Paper. Retrieved on 12 March, 2007 from http://www.w3.org/ECommerce/interface.html. 109

Hartson, R. and Pyla,P. (2018). *The UX Book: Agile UX Design for a Quality User Experience*. US: Morgan Kaufman. xxi, 18

Hayes-Roth, B., Hayes-Roth, F., Rosenschein, S., and Cammarata, S. (1988). Modeling planning as an incremental, opportunistic process. In Engelmore, R. and Morgan, A. (Eds.), *Blackboard Systems*, Addison-Wesley, 231-583. 35

Hill, B. (2010). Diagnosing co-ordination problems in the emergency management response to disasters. *Interacting with Computers*, 22(1): 43-55. DOI: 10.1016/j.intcom.2009.11.003. 3, 177

BIBLIOGRAPHY 197

Hill, B., Long, J., Smith, W., and Whitefield, A. (1995). A model of medical reception—the planning and control of multiple task work. *Applied Cognitive Psychology*, (9), S81-S114. DOI: 10.1002/acp.2350090707. 84

John, B. and Gray, W. (1995). CPM-GOMS: An analysis method for tasks with parallel activities. In *Conference Companion on Human Factors in Computing Systems CHI'95*, ACM. DOI: 10.1145/223355.223738. 5

Kalakota,R and Whinston, A. (1996). *Frontiers of Electronic Commerce*. US: Addison Wesley Longman Publishing Co., Inc. DOI: 10.1109/TCPMC.1996.507151. 106

Kim, G. (2020). *Human-Computer Interaction - Fundamentals and Practice*. UK: CRC Press, Taylor and Francis. xxi, 18

Kirsh, D. (2001). The context of work. *HCI*, 6(2): 306–322. DOI: 10.1207/S15327051HCI16234_12. 3

Lim K. and Long, J. (1994). *The MUSE Method for Usability Engineering*. Cambridge, UK: Cambridge University Press. DOI: 10.1017/CBO9780511624230. 3, 5, 18, 19, 20, 30, 41, 55, 67, 84

Linney, J. (1991). *Review of Reactive Planning Literature*. QMW Dept. of Computer Science Technical Report No. 560. 34

Long, J. (2010). Some celebratory reflections on a celebratory HCI festschrift, *Interacting with Computers*, 2(1), 68-71. DOI: 10.1016/j.intcom.2009.11.006.

Long, J. (2021). *Approaches and Frameworks for HCI Research*. Cambridge: Cambridge University Press. DOI: 10.1017/9781108754972. xxii, 1, 2, 88, 180

Long, J. and Brostoff, S. (2004). Validating design knowledge in the home: a successful case-study of dementia care. In Reed, D., Baxter, G., and Blythe, M. (Eds.). *EACE '12*. France: European Association of Cognitive Ergonomics. 5

Long, J. and Dowell, J. (1989). Conceptions of the discipline of HCI: Craft, applied science and engineering. In Sutcliffe, A. and Macaulay, L. (Eds.), *People and Computers V.* Cambridge, UK: Cambridge University Press. 1, 9, 84, 91, 180

Long, J. and Hill, B. (2005). Validating diagnostic design knowledge for air traffic management: A case-study. In Marmaras, N., Kontogiannis, T., and Nathanael, D. (Eds.), *EACE '05*. Greece: European Association of Cognitive Ergonomics. 5

Long, J. and Monk, A. (2002). Applying an engineering framework to telemedical research: a successful case-study. In Khalid, H. and Helander, M. (Eds.), *Proceedings of 7th International Conference on Working with Computers*. Kuala Lumpur, Malaysia. 5

198 BIBLIOGRAPHY

Long, J., Cummaford, S., and Stork, A. (2022, in press). *Towards Engineering Design Principles for HCI*. Switzerland: Springer Nature. xxi

Long, J. and Timmer, P. (2001). Design problems for cognitive ergonomics research: What we can learn from ATM-like micro-worlds. *Le Travail Humain*, 64(3), 197-222. DOI: 10.3917/th.643.0197. 25

Neale, J. and Liebert, R. (1986). *Science and Behavior: An Introduction to Methods of Research*. Chapter 12: External Validity. Prentice-Hall. pp. 254–274. 68

Nielsen, J. (1993). *Usability Engineering*. San Francisco: Morgan Kaufman. DOI: 10.1016/B978-0-08-052029-2.50007-3. 2, 4

Norman, D. (1983). Design principles for human-computer interfaces. In Smith, R., Pew, R., and Janda, A. (Eds.), *Proceedings of CHI 83, Human Factors in Computing Systems Conference*. Boston, MA: ACM. DOI: 10.1145/800045.801571. 4

Norman, D. (1989). *The Psychology of Everyday Things*. NY: Basic Books. 4

Norman, D. (1993). *Things That Make Us Smart*. NY: Diversion Books. 4, 35

Norman, D. (2010). The transmedia design challenge: Technology that is pleasurable and satisfying. *Interactions*, 17(1), 12-15. DOI: 10.1145/1649475.1649478.

Norman, D. (2013). *The Design of Everyday Things* (Revised Ed.). NY: Basic Books. 2, 4

Pew, R. and Mavor, A. (2007). *Human-System Integration in the System Development Process – A New Look*. US: The National Academies Press. 3

Rauterberg, M. (2006). HCI as an engineering discipline: To be or not to be!? *African Journal of Information and Communication Technology*, 2(4), 163-184. DOI: 10.5130/ajict.v2i4.365. 3

Rauterberg M. and Krueger H. (Eds.) (2000). EU Directive 90/270: State-of-the-art in United Kingdom. *IPO Report No. 1236*. Technical University Eindhoven. 4

Ritter, F., Baxter, G., and Churchill, E. (2014). *User-Centred Systems Design—a Brief History in Foundations for Designing User-Centered Systems*, pp. 33-54. Switzerland: Springer Nature. DOI: 10.1007/978-1-4471-5134-0. xxi, 18

Seffah, A. (2015). *Patterns of HCI Design and HCI Design of Patterns*. Switzerland: Springer Nature. DOI: 10.1007/978-3-319-15687-3. 5

Shneiderman, B. (1983). Direct manipulation: A step beyond programming languages. *IEEE Computer*, 16(8), 57. DOI: 10.1109/MC.1983.1654471. 2, 3, 4, 44

Shneiderman, B. (1998). *Designing the User Interface: Strategies for Effective Human-Computer Interaction*, 3rd Edition. Reading, MA: Addison-Wesley. 4

Shneiderman, B. (2010). *Designing the User Interface: Strategies for Effective Human-Computer Interaction*, 5th Edition. Reading, MA: Addison-Wesley. 2, 4

Smith, M. and Mosier, J. (1986). Guidelines for designing interface software. *Mitre Corporation Report MTR9240*, Mitre Corporation. DOI: 10.21236/ADA177198. 44

Smith, W., Hill, B., Long, J., and Whitefield, A. (1997). A design-oriented framework for modelling the planning and control of multiple task work in secretarial office administration. *Behaviour & Information Technology*, 16(3), 289-309. DOI: 10.1080/014492997119897. 108

Stork, A. (1999). Towards engineering principles for human-computer interaction (domestic energy planning and control). Unpublished Ph.D. Thesis, University of London. xxii, 4, 5, 6, 10, 11, 18, 20, 25, 26, 30, 34, 37, 38, 39, 80, 87

Stork, A. (1992). A Formal Description of Worksystem Behaviours and Interactions. M.Sc (Ergonomics) Thesis. University of London. 86

Stork, A. and Long, J.(1994). A specific planning and control design problem in the home: Rationale and a case study. In *Proceedings of the International Working Conference on Home-Oriented Informatics, Telematics and Automation.* University of Copenhagen, Denmark, 419-428. 96

Stork, A. and Long, J. (1998). Strategies for developing substantive engineering principles. May, J., Siddiqi, J. and Wilkinson, J. (Eds.), *HCI '98 Conference Companion*, 36-37. 83, 96

Stork, A., Lambie, T., and Long, J. (1998). Cognitive engineering coordination in emergency management training. May, J., Siddiqi, J., and Wilkinson, J. (Eds.), *HCI '98 Conference Companion*, 36-37. 83

Teo, L. and John, B. (2008). CogTool-Explorer: Towards a tool for predicting user interaction. In *Proceedings CHI EA'08*, ACM, pp. 2793–2798. DOI: 10.1145/1358628.1358763. 5

Timmer, P. (1999). Expression of operator planning horizons: A cognitive engineering approach. Unpublished Ph.D. Thesis, University of London. 25

Timmer, P. and Long, J. (2002). Expressing the effectiveness of planning horizons. *Le Travail Humain*, 65(2), 103-126. DOI: 10.3917/th.652.0103. 5, 25

Wickens, D. (1984). *Engineering Psychology and Human Performance*. Columbus: Merrill. 2, 4

Wickens, D. (1993). Cognitive factors in display design. *Journal of the Washington Academy of Sciences*, 83(4): 179-201. 2

Wickens, C., Lee, J., and Becker, G. (2004). *An Introduction to Human Factors Engineering*, 2nd edition, Pearson 2004, Chapter 8. 2

200 BIBLIOGRAPHY

Wright, P., Fields, R., and Harrison, M. (2000). Analysing human-computer interaction as distributed cognition: The resources model. *Human Computer Interaction*, 51(1), 1–41. DOI: 10.1207/S15327051HCI1501_01. 3

Zagalo, N. (2020). *Engagement Design: Designing for Interaction Motivations*. Switzerland: Springer Nature. DOI: 10.1007/978-3-030-37085-5. xxi, 18

Authors' Biographies

John Long

University Degrees

M.A. Modern Languages (Cambridge), B.Sc. Psychology (Hull), Ph.D. (Cambridge) and D.Sc. (London).

Books/Theses

Multidimensional Signal Recognition: Reduced Efficiency and Process Interaction (Ph.D.), *Attention and Performance IX*, with Alan Baddeley (LEA), *Cognitive Ergonomics and Human-Computer Interaction*, with Andy Whitefield (CUP), *The MUSE Method for Usability Engineering*, with Kee Yong Lim (CUP), *Approaches and Frameworks for HCI Research*, (CUP), and *HCI Design Knowledge – Critique, Challenge and a Way Forward*, with Steve Cummaford and Adam Stork (M&C).

Steve Cummaford

University Degrees

B.A. Philosophy (York), M.Sc. Cognitive Science (Cardiff), Ph.D. (London).

Books/Theses

The Effects of Expected and Unexpected Interruptions on Completion of Computer-based Tasks (M.Sc. Thesis), HCI Engineering Design Principles: Acquisition of Class-Level Knowledge (Ph.D.), and *HCI Design Knowledge: Critique, Challenge and a Way Forward*, with John Long and Adam Stork (M&C).

Current Position

Lead digital product designer at Ted Baker

Adam Stork

University Degrees

B.Sc. Computing and Robotics (Kent), M.Sc. Ergonomics/Human-Computer Interaction (London), and Ph.D. (London).

Books/Theses

A Formal Description of Worksystem Behaviours and Interactions (M.Sc. Thesis), Towards Engineering Principles for Human-Computer Interaction (Ph.D.), and

202 AUTHORS' BIOGRAPHIES

HCI Design Knowledge: Critique, Challenge, and a Way Forward, with John Long and Steve Cummaford (M&C).

Current Position

Partner, also strategy and transformation consultant at Concerto.

Lightning Source UK Ltd.
Milton Keynes UK
UKHW020802150422
401586UK00003B/48